Everyday Applied Geophysics 1

Series Editor
André Mariotti

Everyday Applied Geophysics 1

Electrical Methods

Nicolas Florsch
Frédéric Muhlach

ISTE Press Ltd
27-37 St George's Road
London SW19 4EU
UK

www.iste.co.uk

Elsevier Ltd
The Boulevard, Langford Lane
Kidlington, Oxford, OX5 1GB
UK

www.elsevier.com

Notices
Knowledge and best practice in this field are constantly changing. As new research and experience broaden our understanding, changes in research methods, professional practices, or medical treatment may become necessary.

Practitioners and researchers must always rely on their own experience and knowledge in evaluating and using any information, methods, compounds, or experiments described herein. In using such information or methods they should be mindful of their own safety and the safety of others, including parties for whom they have a professional responsibility.

To the fullest extent of the law, neither the Publisher nor the authors, contributors, or editors, assume any liability for any injury and/or damage to persons or property as a matter of products liability, negligence or otherwise, or from any use or operation of any methods, products, instructions, or ideas contained in the material herein.

For information on all our publications visit our website at http://store.elsevier.com/

British Library Cataloguing-in-Publication Data
A CIP record for this book is available from the British Library
Library of Congress Cataloging in Publication Data
A catalog record for this book is available from the Library of Congress
ISBN 978-1-78548-199-4

Printed and bound in the UK and US

Contents

Foreword

The scientific books published by ISTE Press include a multidisciplinary series called *Earth Systems – Environmental Sciences*, and it is in this context that today I present a work dedicated to geophysical prospecting and its applications, coordinated by Professor Nicolas Florsch.

Its title, *Everyday Applied Geophysics*, deserves to be explained in more detail.

First, we should recall the important role played in some scientific fields by the so-called "amateurs". This is especially the case for astronomy, a field where a socioepistemology of amateur practices, whose main points can be summed up here, has been established. These amateurs are not organized to compete with professionals, as they evidently lack the skills and the necessary resources. However, this is not a case of popular science: their practices, beyond the understanding of the sky, stars and the universe, are active and mobilized by the desire to make discoveries. Astronomy is a science where amateurs can obtain significant observation data, which are very useful for scientists.

On a smaller scale, some amateurs, for example, are quite involved in electronics and radio communication.

However, so far this has not been the case for Everyday Applied Geophysics, a domain that has potentially numerous applications associated with the exploration of the near subsoil: looking for water, archeological remains, geological peculiarities, etc.

Moreover, making Everyday Applied Geophysics available for researchers based in developing countries is a challenge of the utmost importance.

The goal is to open this field and allow everyone to employ the tools and methods used for exploring the near subsoil in order to highlight reservoirs or flow paths, locate holes, define geological stratifications, follow pollution plumes, search for archeological remains, etc. If curious and exploring amateurs may be involved, the main objective of the scientific community of these countries, which needs financially and technologically affordable tools, is to implement cheap and unsophisticated methods and techniques that, nonetheless, will produce plenty of essential data.

Let us provide an example to illustrate this point. Some electrical tomography devices cost tens of thousands of dollars on the market; in this work, we will discover that with €100 at most, we can implement a system that, despite being naturally three to five times slower in terms of data acquisition, allows everyone to carry out actual and effective electrical tomography.

This work will also focus on the issue of self-learning. The existing literature does not tackle practical aspects either in terms of material implementation or basic interpretation concepts (actual resolution of the methods, sensitivity, etc.). This work is also very useful insofar as it can solve the problem of signal acquisition: it provides open-source "Arduino" solutions, supported by a downloadable program, for data acquisition in the field.

Thus, this work, which is unique in its genre and accessible to everyone (with a few more technical and/or mathematical boxed passages), bridges a double gap in the existing scientific literature by:

– providing accessible tools for the exploration of the near subsoil: from tools to acquisition systems (the latter being available with the use of computers) including a guide of free programs;

– providing practical information for implementation that cannot be found in other works, such as the design of devices (from electrodes to current flow, for example to carry out an electrical survey), the protocol for the creation of geophysical maps, etc.

We hope that this work reaches its audience and that the scientists that played a part in it may thus contribute to the removal of the ideological barrier between the world of basic research carried out in the academic world and applied research, as the markedly ideological gap that divides these two communities has not been entirely bridged yet. Besides, helping the development of the environmental field should be invaluable for a large number of countries.

André MARIOTTI
Emeritus Professor at the Université Pierre-et-Marie-Curie
Honorary member of the Institut Universitaire de France

Introduction and General Points

1.1. Introduction: the audience of this book

This book is dedicated to curious individuals and explorers of knowledge and technologies that have not been awarded a Nobel prize (yet). It has been written for tinkerers who want to employ tools to analyze the near subsoil (less than 20 m) with what is at hand. It has been designed for all those who want to carry out what we call "applied" geophysics (or "subsurface geophysics" as well as "environmental geophysics", etc.). The authors have mentioned it with individuals from modest backgrounds in mind, as geophysical materials are expensive (naturally, as they are not produced in the thousands).

If we had to point out a prerequisite, it would be a certain familiarity with the laws of electricity and the difference between tension (or difference in potential as well as voltage, in volt) and intensity (in amperes). Thus, neophytes and amateurs of all kinds are welcome.

Sometimes, more complex calculations are shown in gray boxes for those math brains with a more developed interest, but they do not constitute the bulk of the work and may be disregarded without any serious consequence.

This book assumes that the reader will rummage on the Internet, looking for the all possible additional information at his or her level: figures, documents of all kinds, we can even find some videos dedicated to geophysics on YouTube. The content of this work can hardly be found in or is completely absent from specialized books, which besides are very

The authors would like to thank Michel Kammenthaler and Christian Camerlynck for their contribution to this book.

expensive, and/or Websites dedicated to applied geophysics and its applications. We propose opening new avenues and offering some practice. The authors rely on the reader's curiosity and are convinced that increasing one's personal experience through practice is a fulfilling activity.

One of the goals is that the reader, after reading this book, will be able to analyze the subsurface and carry out electrical surveys to find out the depth of an aquifer or electrical mapping to find out the clay portions of the soil.

There is relatively little space in this first volume for applications: a choice had to be made. However, several examples of applications can naturally be found on the Internet.

1.1.1. *What is applied (or subsurface) geophysics?*

Applied geophysics is the use of concepts, quantities, devices and physical sciences to explore the near subsurface. Think about it… with our own eyes we can see the Andromeda Galaxy (on a moonless and cloudless night, without any interference lights from cities and provided that we are in the northern hemisphere; in the southern hemisphere, we will see the Magellanic Clouds, i.e. galaxies orbiting the Milky Way). Andromeda is 2 million light years away, i.e. around 24 trillion km. However, in our garden, there is no way of *seeing* what goes on below the first millimeter of soil.

There is still time to close this book and move on to astronomy.

Our "remote senses", such as sight, require physical quantities (in this case light), which do not cross the soil. If only the soil were as transparent as the crystalline water of a lagoon.

Applied geophysics precisely complements our senses, simultaneously replacing our organs with instruments and quantities with other quantities to which these instruments are sensitive and which are above all able to *penetrate* the soil. The goal of geophysics is exactly to make the soil transparent. Our eyes do not see these physical fields[1], such as the electric field or the magnetic field. The idea here is to transform these fields,

---

1 An alfalfa field is a space covered all over by alfalfa. In a magnetic field, we replace the alfalfa with a force that orients the needle of the compass and exists everywhere. Similarly, we see the action of an electric field in any point of the space on an electric charge.

measured with instruments, into graphs and maps that this time will be interpretable. To begin with, let us consider three examples. First, let us consider electric current. It crosses the materials that form the subsoil fairly easily – otherwise, where would lightning go and what is the purpose of earthed sockets? However, these materials can be more or less conducting. A clay soil, for example, is easily a thousand times more conducting than a granite soil. Second, magnetic "fields", the very ones that orient our compass: they cross the soil fairly well, unless they find a metallic barrier (for example heavily reinforced concrete). Third, acoustic waves; it is better to refer to seismic waves, but it is the same kind of physical process: a mechanic perturbation propagates in space.

Applied geophysics aims to see what goes on underground, let us say from a few centimeters to a few dozen meters, unless we focus on oil exploration, which goes deeper. Beyond this, the means must increase, and we will refer instead to global geophysics.

"Seeing" matter is something that medical imaging can also perform with X-rays and a scanner, magnetic resonance imaging, and several other methods.

Just like the brain interprets the signals that trigger our senses, an (applied) geophysicist must then interpret (by reasoning, but also by relying on their cumulative skills and experience) "data", "observations" and "measurements" to deduce something, if possible a map of the subsoil.

To deduce what exactly? We must be realistic. In Jurassic Park, paleontologists "image" a small dinosaur skeleton on a screen. However, there are some physical limitations, which we will come back to later, preventing us from obtaining the same resolution as our eyes.

In terms of demand, in any case, applied geophysics helps us find and describe aquifers. It can map soil parameters whose subsoil nature and geometry may be determined, using some supplementary information (for example the geological context), in the sector explored. To a certain extent, it can determine the position of a fault or the thickness of a clay layer, as well as many other things, for example it can mark out polluted areas, find buried metallic objects or outline the foundations of former settlements.

Applied geophysics is in no sense a universal solution. It is more precise and exact to say that it spells out what the subsoil is not rather than what it is. It proceeds by refuting and eliminating. This is due to the fact that physical laws are inflexible and do not agree on their own with our thirst for knowledge. Those who do not understand this will hit a dead end. In the end, in the face of the great variety of geological and environmental situations, geophysics only provides some physical parameters, such as resistivity or the speed of seismic waves. An infinitely rich world in the subsurface is met by a very limited geophysical knowledge.

What is worse, the size of the solution spaces may be huge. What does this mean? Let us consider the following problem: find two numbers whose sum is 10. The solution is $2 + 8$, $5 + 5$, as well as $-35 + 45$ and even 82, $745 + (-82,735)$. This is often the dilemma faced by geophysicists, and even what constitutes their core business: extracting from the infinite space of solutions the most suitable solution according to a specific criterion, which will sometimes be more or less meaningful. The aforementioned numerical example clearly illustrates this point. The information that the sum is equal to 10 is actually significant: think about all the cases eliminated. On a Cartesian plane, it is a straight line $x + y = 10$, instead of all the space of (x, y). The reduction of the set of possible points is considerable[2].

1.1.2. *The spirit of this book: affordable methods on a technological and financial level*

Supply and demand feed off each other and develop the use of geophysics. Despite everything, applications remain confidential in the sense of being uncommon in society (unlike smartphones or cars). There is no mass dissemination, so that the increased performances of the tools are coupled with prices that do not go down. Does this mean, however, that geophysics is only accessible for a lucky minority? This book illustrates the opposite.

It turns out that some techniques remain very similar to basic methods, often very simple, and that only the approaches have become more complex,

---

2 For geophysicists, but not for mathematicians, showing that the plane and a straight line have the same, infinite number of elements. This is due to the fact that the geophysicist's straight line has a certain thickness and a thin length, with limited resolution; in short, a limited number of degrees of freedom.

especially to save time (and the time...). Other methods have involved miniaturization, especially sensors, and some steps that were quite hard to take in the past have become easy to implement.

The goal of this book is to make some techniques used in the professional world of applied geophysics accessible to everyone; not all of them, even though we may suppose that this approach will become more widespread in the near future, especially with the popularization of all sorts of sensors and the ever-increasing ability to digitize and store signals.

The bibliographic issue is tricky. Naturally, all the points tackled in this book are developed in the scientific literature. However, the articles of specialized journals are nearly always in English and accessible only for a fee. Money must be paid before seeing the article on the basis of a short abstract that is to the content of the article what a label is to a tin can.... We can never be sure that an article that cost tens of dollars will meet our expectations. For professionals, it is then the employer's organization, whether a university or a research body, that pays a subscription, so that there is no problem if a paper is discarded after being read and considered uninteresting. However, for a merely curious individual, this is a lot to pay.

The Internet allows us to find what we want based on a few keywords. Try to type "electrical tomography" on Google – it is a simple example – and several links and/or images pop up. Naturally, it is necessary to make a selection, but is that not part of the discovery?

Thus, we have chosen bare-bones bibliography that can be accessed online. To find out more, readers may use "Google Scholar", where several specialized articles can be downloaded.

The (#) symbol is an invitation to carry out research online and in particular on Wikipedia. This (#) confers to the volume a scope that is incredibly broader than it seems at first: it is always an invitation to set out on a scientific journey.

This book aims to be simultaneously elementary and practical. Through a bibliography that can be accessed online and a list of developed programs that can be found online (throughout the text or in the last chapter), it does not overlook complexity.

1.1.3. *An example to begin*

Let us consider a power generator, for example some batteries or – even better – a small alternating current generator like the one described in section 5.3 dedicated to the instruments employed for electrical prospecting. Let us design the follow system: the voltage produced by the generator is carried by two wires toward two electrodes, simple metallic posts planted in the ground, called C_1 and C_2. We connect a voltmeter to another two electrodes called P_1 and P_2. We create a given geometrical structure: the electrodes are separated (at least 20 m apart, for reference), except for C_1 and P_1, which stand 1 m apart. Figure 1.1 illustrates this system, where a wooden stand mechanically connects C_1 and P_1. We note a = C_1P_1. The device used in our example is represented in Figure 1.2.

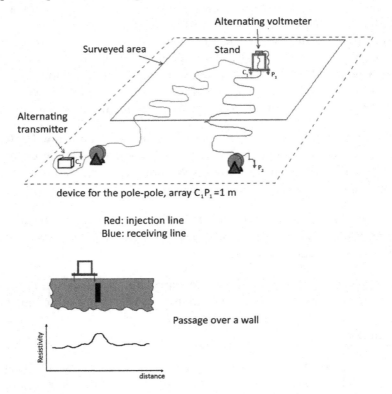

Figure 1.1. *Pole–pole electric device used in this example. The stand is moved to explore the terrain, as the measurement is related to its close environment. For a color version of this figure, see www.iste.co.uk/florsch/geophysics1.zip*

Figure 1.2. *A picture of the system used at this site, with its data acquisition device*

Let us consider the quantity ρ_a (rho index a) called apparent resistivity (expressed in $\Omega.\text{m}$), given by:

$$\rho_a = 2\pi a \frac{\Delta V}{I},$$

where ΔV is the potential difference measured by the voltmeter and I is the power measured by the ammeter. Apparent resistivity is representative of the tendency of the soil to resist the current flow. We can see at the bottom of Figure 1.1 how the presence of a wall (*a priori* more electrically resistant than a ground enclosure, for example) contributes to an increase in apparent resistivity. The measurement is related to the center of $C_1 P_1$. Provided that these resistant areas are organized, they may reveal the organization of the underlying structures, such as archeological structures.

Moving meter by meter on a square mesh, we obtain a *map of apparent resistivity*, as is shown in the example of Figure 1.3. Overall, this map includes around 10,000 measuring points, equal to a day's work. The method clearly underlines some foundations. Here, they represent several houses of a Gallo–Roman farm in the northeast of France, close to the village of Dehlingen. The positioning mode uses simple measuring tape on the ground that marks the position of the measuring point. They can be seen on the ground in Figure 1.2.

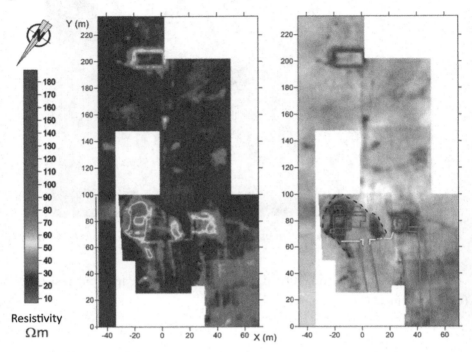

Figure 1.3. *Pole–pole electrical map obtained at the Gurtelbach site and interpreted, on the right, depicting some walls. For a color version of this figure, see www.iste.co.uk/florsch/geophysics1.zip*

1.1.4. *Drawing maps in isovalues*

Drawing maps like the one shown above requires specialized software. We use the commercial program SURFER published by Golden Software, whose trial version can carry out basic operations. Readers may use GIS programs such as QGIS, which is free and powerful, or install an open-source package like Scilab. The advantage of Scilab is that it simultaneously represents an extraordinary new avenue for scientific computing. Another very useful package is "R", which is an open source and large-scale collaborative program dedicated to statistics[3]. It can carry out every

---

3 https://www.r-project.org/. See also https://en.wikipedia.org/wiki/R_(programming_language).

operation required to draw maps, and readers will be able to find several "tutorials" online.

On a general level, two-dimensional data (x, y) are a collection of N points – measurements F:

$$\left\{ \begin{array}{ccc} x_1 & y_1 & F_1 \\ x_2 & y_2 & F_2 \\ x_3 & y_3 & F_3 \\ ... & & \\ ... & & \\ x_N & y_N & F_N \end{array} \right\} .$$

The points (x, y) may or may not be on a regular grid. The first step is to interpolate the values on a regular grid based on the interval $(\Delta x, \Delta y)$, before drawing the map (these programs prefer this type of regular grid).

1.2. Direct method, inverse method

This section may also be read after the chapters that describe the techniques. Some notions become clearer when we have mastered a technique, or even with the first experiences in the field.

In our daily life, several measurements can be taken with instruments that provide an immediate result. This is the case when we measure the length of a simple object with a ruler, or when we test cell voltage with a voltmeter.

In geophysics (and in many other fields!), determining a quantity we are interested in is a more complex process. Eratosthenes taught us a useful lesson by measuring the diameter of the Earth.

Let us consider the example of the depth of the "top" (upper limit or surface) of an aquifer or "water table". The simplest process involves carrying out some drilling and then measuring at which depth we can find water. However, the goal of geophysics is to obtain this result without carrying out any expensive drilling operation…which could also reveal a total lack of water. To obtain this depth, most geophysicists will aim to carry out so-called vertical electrical sounding, which involves sending current to the soil with two so-called transmitter electrodes (C_1 and C_2) and measuring the potential on two other so-called receiving electrodes (P_1 and P_2). This method is described in section 2.10. To this end, we can use four equidistant electrodes that are gradually separated to inject current at an increasing depth. The "measurement" then represents a set of potential and current values $\{V_i, I_i, i = 1...N\}$ associated with a set of interelectrode spacing $\{a_i, i = 1...N\}$, where N is the number of measurements. Figure 1.4 illustrates this principle: the greater the distance between the transmitter electrodes, the deeper the current penetrates.

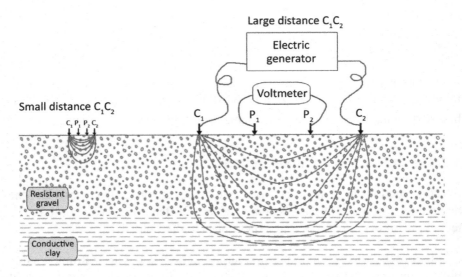

Figure 1.4. *The principle of a resistivity measurement. Resistivity includes the influence of the underlying ground. With larger devices, current penetrates more deeply. Too few measurements provide scarce information about the complexity of the soil, occasionally making the interpretation uncertain. When carrying out electrical sounding, we hypothesize that the soil is "horizontally layered", and the issue is to determine the thickness and resistivity of the layers based on surface measurements. For a color version of this figure, see www.iste.co.uk/florsch/geophysics1.zip*

Thus, we are led to the fundamental difference between two spaces (or sets): the space of measurements, which constitute a certain type of physical quantities (here, voltage V and current I), and the space of parameters (here, the resistivities of the two layers and the depth of the interface between them), which represent the quantities we are actually interested in and cannot in any way deduce "immediately" from the measurements. Moreover, there is another space – which is the most important one – i.e. the space of reality. This is due to the fact that, in the end, underground water is at a given depth, and assuming that the top of the aquifer is situated at a depth "d" means already seriously hypothesizing that the surface of this aquifer is naturally flat and horizontal, that this water reservoir actually exists, that our measurements will be sensitive to it and that we are not mistaking for an aquifer (*a priori* more conducting) what may be a mere impermeable clay layer.

Figure 1.5 illustrates these three spaces. What we *know how* to do due to mathematics (or with a free or commercial program that carries out these calculations) is, given a certain structure, to calculate (predict, simulate) the measurements that would be obtained for this structure. This is the *direct problem*: we are given the parameters, we deduce the "measurements that we would obtain" once these parameters have been established.

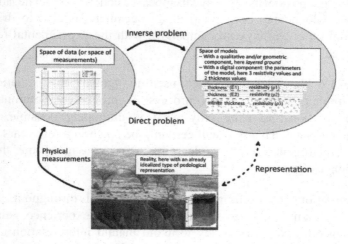

Figure 1.5. *The general paradigm of geophysics, where the issue is always to transform measurements into model parameter assessments, which are our representations of a hardly accessible reality (at least without excavators). For a color version of this figure, see www.iste.co.uk/florsch/geophysics1.zip*

The *inverse problem* represents the opposite approach: given a set of measurements taken during an experiment, what can we say about the structure and the parameters? It is this situation that interests us in the first instance!

The parameters are those of a *model*: we can define quantitative parameters (to be determined) only if they are related to something, especially a type of preestablished geometry. In the aforementioned example, the model may involve a kind of soil made of alluvial deposits, such as gravel, where water can flow freely without any significant lateral movement, so that the name "water table" is appropriate: it is more or less horizontal. In similar cases, the complete model includes a geometric and geological choice (an interface with the water-saturated environment in homogeneous gravel) as well as quantitative parameters, for example the depth of this water table.

Dealing with the inverse problem is the core business of geophysicists. As shown in the example provided in the Introduction, is 10 equal to 2 + 8 or 575 − 565? This vagueness is not a simple understatement; it is normal business for geophysicists. It may disappear when we associate additional information, which may be geological, geotechnical, related to drilling or derived from other geophysical campaigns, with the experimental results to be interpreted.

Figure 1.6 shows the approach used to "grasp" this situation. The practical implementation of this process of progressive elimination of unsuitable solutions, to keep only what is ultimately possible, is very difficult to automate. This step will certainly be taken in a few years because of artificial intelligence and "deep learning" (#). In the meantime, this is the job of geophysicists.

The limitations of each method are mentioned throughout the text. However, it is impossible to list all of them: the experience gained and common sense are always the key to avoid blatant interpretation mistakes. Let us emphasize this point: the correct way of seeing geophysical analysis involves the observation that geophysics provides more information in the shape of eliminated possible cases than in the form of positive claims. Measurements show what the subsurface is not, they eliminate a large part of possible solutions, and this "result" must be combined with other information to remove ambiguities.

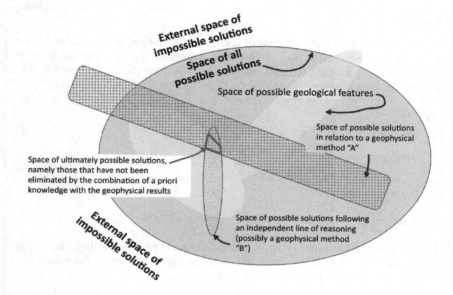

Figure 1.6. *The way in which information is used in geophysics and several other sciences. The issue is that nearly everything is possible when, at first, our knowledge is limited. Gathering the available information involves the gradual elimination of those cases that are not compatible with the observations. For a color version of this figure, see www.iste.co.uk/florsch/geophysics1.zip*

There are several cases where no mathematical inversion is necessary or explicitly implemented, as it is the geophysicist that carries out this operation mentally. In the first example of the Gallo-Roman villa, no digital inversion was carried out, as the image was sufficient. However, the instinctive reaction of geophysicists as well as archeologists is to attempt to transform this map into quantified structures, i.e. a set of walls whose thickness is estimated. They move *ipso facto* from the space of measurements (apparent resistivity values) to a space of parameters (nature of the material, earth or stone, width, length and position of the walls). We may say that they carry out a qualitative inversion because of their neurons, which carry out the necessary operations.

How does "inversion" work?

The principal method is the method of least squares. In the classic problem of the regression line, we need to find two numbers a and b so that, considering a set of N points (x_i, y_i) belonging to the plane Oxy, we can draw the line $y = ax + b$. To this end, for each point we should obtain that:

$$\begin{cases} y_1 = ax_1 + b \\ y_2 = ax_2 + b \\ y_3 = ax_3 + b \\ .... \\ .. \\ y_N = ax_N + b \end{cases}$$

The only thing is that some experimental points are not in line so that, except for 2 points (always touched by a line), the equation does not apply to the whole of the system. Thus, we should lower our expectations.

If the equation does not apply, let us try at least to make the differences between (y_i) and $(ax_i + b)$ "as small as possible". How can we make a set of numbers "as small as possible"? The idea is to see the differences $\left[y_i - (ax_i + b) \right]$ as the components of an "error" vector, so that $\varepsilon_i = \left[y_i - (ax_i + b) \right]$. This vector must be as small as possible, and what its magnitude represents must be as small as possible. Because of the good old Pythagorean theorem (with N dimensions!), this magnitude is:

$$\mathrm{mod}(\vec{\varepsilon}) = \sqrt{\sum_i \varepsilon_i^2} = \sqrt{\sum_{i=1}^{N} \left[y_i - (ax_i + b) \right]^2}$$

To make calculations easier, we prefer using – there is no difference – the square of this magnitude, here noted as J:

$$J(a,b) = \sum_{i=1}^{N} \left[y_i - (ax_i + b) \right]^2$$

Thus, we now face a new problem: finding (a,b) so that J is as small as possible. This is the "problem of least squares".

This book does not aim to develop these mathematical techniques, and there are numerous tutorials and videos about the *method of least squares*(#).

If we analyze in more detail the problem of the line, we realize that $\{ y_i; i = 1, 2...N \}$ represents the data (d) of the data space (the x values are assumed to be known and therefore do not have to be found) and (a,b) belongs to the space of the parameters: this is what we are looking for. We moved from a problem in the data space (drawing the best line possible) to a problem in the parameter space (finding the best a and b).

The "model" represents the choice of a line, and it is associated with specific parameters. Usually, we also refer to a model to designate the set: the type of structure (here, the line) *with* its parameters.

Let us go back to geophysics and extrapolate. Let us note the N data as $\vec{d} = (d_1, d_2, ...d_N)$ and the M parameters as $\vec{p} = (p_1, p_2, ...p_M)$. In the example of the line $\vec{p} = (a, b)$, the direct problem shows that the observable elements (the data) are a function, simple or complex, of the parameters of the model, which may be written as:

$\vec{d} = \vec{G}(\vec{p})$. Component by component, with the example of the line (M=2):

$$\begin{cases} d_1 = G_1(a, b) = G_1(p_1, p_2) \\ d_2 = G_2(p_1, p_2) \\ \text{etc...} \\ .... \\ d_N = G_N(p_1, p_2) \end{cases}$$

For the line: $G_i(a, b) = ax_i + b$.

The problem of least squares can then be written as:

Find $\vec{p} = (p_1, p_2, ...p_M)$ so that $J = \sum_{i=1}^{N} [d_i - G_i(\vec{p})]^2$ is as small as possible.

The rest is an exciting mathematical adventure, like the one offered by Albert Tarantola on his Website[4]. If inversion, as seen in such a document, seems noticeably mathematical, interpretation programs, especially those mentioned in this book, are fairly easy to use and do not require actual math skills.

With more or less significant subtleties, the vast majority of inversions in geophysics employ the least squares method. This will apply to all the inversions mentioned in this book.

---

4 https://www.ipgp/fr/~tarantola/. See in particular "Generalized Nonlinear Inverse Problems Solved using the Least Squares Criterion". The reader will have to be familiar with matrix calculation.

1.3. Sampling quantities on the surface, and which resolution can be reached underground?

1.3.1. *General points about measuring protocols*

Applied geophysics involves one-, two- or three-dimensional (one, two, or three spatial variables – 1D, 2D, 3D) measurements in the physical space. Some methods and/or objectives require a temporal component, creating then what we call 4D[5]. First, we face the issue of finding out which measuring scale or grid size to consider. We also refer to 2.5D (or 2D1/2) when the structure is 2D but the source is 3D (especially for point electrodes in 2D tomography).

There are two common 2D approaches. One involves maps related to a system of ground coordinates (0,x,y) (if we considered a continent, we would prefer using latitude and longitude), while the other involves sections, which are in a sense vertical maps, exploring along a surface x-axis (x) and in depth (z). We can only refer to actual depth after inverting. In the space of parameters, raw measurements may be represented on a vertical axis, but this would be only one way of representing the data called *pseudo-section*.

Thus, although electrical sounding has a local value (1D with depth z), sounding several times by displacing the center along an axis (x) will allow us to carry out a 2D pseudo-section.

Vertical 2D images are often called *electrical resistivity tomography*.

If we consider three dimensions, we associate a surface exploration in relation to (0,x,y) and z to put forward interpretations and models according to (0,x,y,z). 3D analysis is more time consuming and expensive than 2D and consequently 1D analysis. However, it may provide useful information about structural geometry that a 2D analysis would not reveal.

In tomography, 2D is often enough, in the hope that the underlying structures, at least at the maximum distance to which the device is sensitive, resemble a 2D reality, i.e. markedly stretched-out structures on the plane perpendicular to the section. We will make the most of our knowledge of the geology of the soil before determining the sections. For example, if we

---

5 These dimensional variables play the same roles as the coordinates (x_i) of the line. They are not part of the quantities that have to be inverted.

wanted to explore the arrangement of the layers of an alluvial valley (gravel, sand, clay), it would be natural to create sections or pseudo-sections perpendicular to the axis of the valley. This is also a rule of thumb: whenever possible, we should prospect perpendicularly to the principal length of the expected structures.

However, if we know nothing about the underlying structures, carrying out a 2D analysis when we are standing on a markedly 3D structure (for example a circular sinkhole hidden by a clay cover) carries a risk (of making a mistake). Thus, we can consider a simple form of control: carrying out *two* additional "control" sections, which is common practice. However, this does not reveal a 2D structure, as there are simple layouts whose soundings or sections will be identical, as is the case for an analysis based on a circular structure.

If two perpendicular soundings differ, we know that we cannot confirm the presence of a horizontally layered layout. As for the use of 2D sections, we need to create two parallel sections to confirm or reject the 2D nature of the underlying structures. In this and several other cases, it is refutations that are certain, and the Internet is unfortunately full of examples of interpreters who "see" detailed geological features[6]. There are numberless cases, and it is *always* dangerous to be categorical in our conclusions. Thus, if an electrical sounding is centered on a sinkhole, the result will be the same regardless of the orientation of the sounding, whereas we are far from a horizontally layered layout. This situation may convince a slightly hasty geophysicist that he is standing on layered ground, whereas this is certainly not the case. As we have already pointed out, geophysical measurements correctly tell us what the subsoil *is not* rather than what it is[7]. It cannot be repeated often enough! In this case, we will draw the correct conclusion by saying that the subsoil is not asymmetrical in relation to the center of the sounding but, that being said, we should not make the opposite mistake. Taking risks is part and parcel of the interpretation and takes into consideration probabilities that can even be subjective. Otherwise, our knowledge about the soil would never increase...

---

6 Everyone can buy a device, take some measurements and run commercial inversion programs. This is not what geophysicists do, but that does not prevent people from selling a service.

7 We cannot help but make a comparison with the definition of science given by Karl Popper(#), who qualifies as scientific what is *refutable*.

1.3.2. *The issue of the measuring grid*

It is difficult to establish a general rule, and it is often a matter of finding a compromise in the scales. To choose a "good" measuring scale, we should take into consideration different factors. Let us list some of them.

1.3.2.1. *First: the method*

In Galvanic methods[8], if we note the distance between the observation point and the structure affecting the measurement as r, the resolution is in the range of $1/r$ before inversion, and after a well-conducted inversion it may range from $1/(2r)$ to $1/(5r)$. This should not come as a surprise: the resolution (both lateral and vertical) will never be more than 2 or 3 m for a target situated 10 m underground. There is no point in expecting to see any detailed feature[9]. The only exception is if we make strong hypotheses about the model – for example, assuming during an electrical sounding that we are standing on *horizontally layered ground* – in which case it is possible to reach 5% of uncertainty at best (once again, if this hypothesis is valid!). Often, two cross soundings are carried out, but it has been noticed that this does not represent a totally distinguishing factor.

At the other end of the resolution spectrum we can find the methods that rely on waves: georadars(#) and seismic reflection. Due to the complexity of the instruments, these methods will not be tackled here. It should be pointed out that seismic reflection cannot be (yet) used for the near surface (less than 5 or 10 m), whereas radars do not work in clay areas and their structural interpretation is tricky[10]. The resolution of electromagnetic methods such as magnetotellurics(#) occupies an intermediate position in relation to the two other methods.

---

8 In geophysics, "galvanic" is a synonym of direct current. Measurements are not sensitive to a potential *electrical polarization*(#) and frequencies low enough not to create *displacement current*(#).

9 Let us take two ordinary light bulbs and put them under a frosted glass or some tracing paper. If we look at them from above, we will clearly see two patches of light. Let us move the lamps away from the frosted glass and downwards. We will quickly see the two patches of light merge into one (unless the type of light is very directional, as is the case for a laser beam, naturally). This is similar to the case of direct current electrical methods.

10 However, radars as tools for the *detection* of underground structures such as pipes is effective and commonly used, as the images obtained are very revealing.

1.3.2.2. *Second: the depth of investigation and the representative volume "seen" by devices*

If an object is targeted at a given depth h, on the one hand, the soil must be transparent to the field used to reach it and, on the other hand, the influence of this target must actually reach the surface where the measurements are taken (assuming as a minimum that the target is not too small).

In the electrical example considered in the Introduction, the depth of investigation is linked to the distance a = C_1P_1 and considered to be *in the range of* a/3 or a/2. Thus, a tool with a = 1 m, which was used here, is suitable for a wall body situated at a depth of 50 cm (with a surface deeper than 30 cm). In practice, the depth of investigation depends on several parameters, and it is tricky to provide a universal definition. There are one or two "official" definitions, such as the depth at which, for a given device, a thin layer would be the most visible when passing at this depth. Another definition, namely the "effective depth of investigation", is the depth beyond which the soil has the same influence as the soil contained within it. This seems a good definition, but it "forgets" that sensitivity, with depth, is not at all symmetrical to this boundary. An object buried twice or three times deeper will mostly be detected. In short, the penetration of current depends on the structure of the environment, and if a definition of the depth of investigation is meaningful in itself, it will hardly help us. Thus, let us imagine a (resisting) wall encased in a conducting rock (clay), both situated under a resisting cover (embankment). Then, the current will "look for" the conductor in depth, and the target will be clearly detectable. Let us invert the example: a ditch dug in a resisting rock, filled with much more resisting materials, both lying under a clay covering. In this case, the covering will channel the current on the surface, and it will be hard to detect the ditch since hardly any of the current will have crossed it, remaining in the superficial layer. Nonetheless, the geometry of the structures is the same. To conclude, let us say that we relate the depth of investigation to the device used, but this does not imply that we reach said depth while prospecting. It is better to regard this notion as vague: we will avoid contradictions. It plays a role in a relative line of reasoning, for example when we say that a given device "can see" deeper than another… but that depends once again on the structure of the soil[11]!

The representative volume may be defined as the part of the ground that influences a measurement. If we consider a pole–pole device, such as the one used in the example provided in section 1.1.4, it constitutes a volume that horizontally encompasses the electrodes and is slightly deeper than the depth of investigation. In this case, it is a sort of slightly elongated half-potato.

It would be possible to give a rigorous definition of the representative volume – some definitions exist – but since the direction of the current in reality depends on the conformation of the subsoil, its usefulness is quite dubious.

The only notion that is actually useful in these discussions about depth of investigation and representative volume is that of *a posteriori* sensitivity, and modern programs (such as RES2DINV) illustrate this quantity on a section. Let us consider a small representative volume in the inverted image: we face the issue of finding out if it really affects the measurement, *taking into consideration the results of the inversion, and therefore the actual structures*. If J represents the sum of the squared deviations during the inverse process (see the box about inversion), then it represents the map of $\frac{\delta J}{\delta \rho}$, where δJ is the variation of this J "cost" function resulting from a variation $\delta \rho$ in the small volume considered. *This constitutes a significant "output" of the inversion, since in those cases where sensitivity is low, the results are not meaningful: they are not constrained by the measurements.*

1.3.2.3. Third: the wavelengths or sizes of the expected anomalies

If we consider a direct current electrical method, for structures with a top situated at a depth z, we should expect surface wavelengths of the same order of magnitude as z. Considering a scale greater or even equal to z entails the risk of "missing" the anomaly and being unable to represent its actual shape (Figure 1.7). It is wise to choose a scale smaller than z. In practice, z/2 to z is a compromise. A real one! Let us imagine that we have

---

11 https://www.eoas.ubc.ca/ubcgif/pubs/papers/geop64-403.pdf represents an example of a specialized article focusing on the issue of the depth of investigation.

to prospect in 2D to find the foundations of a house situated 50 cm underground, with walls that are larger than 1 m. If we cover 1 hectare following a 1 m × 1 m grid, we obtain 101 × 101 = 10,210 points. At a rate of 1 point/2 s, which is rapid in this field, it will take us 6 h.

If we moved on to a 50 cm × 50 cm grid, this would quadruple the number of points and the measuring time. On the other hand, with a 2 m × 2 m grid, we may miss the anomaly.

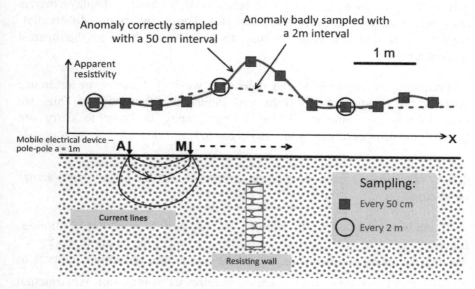

Figure 1.7. *In this example of electrical prospecting, an excessively large sampling interval (2 m in this case) makes us miss the information we are looking for. A 50 cm interval, on the other hand, is appropriate. For a color version of this figure, see www.iste.co.uk/florsch/geophysics1.zip*

Therefore, we can add the fourth point.

1.3.2.4. *Fourth: while prospecting, we are looking for a resolution duration trade-off*

If we want a simple rule, we should sample at an interval that is *at most* of the same order of magnitude as the expected depth. However, this rule remains vague.

1.3.3. *The question of georeferencing*

Prospecting with measuring points that are not related to a system of coordinates, preferably stable, means not prospecting at all. What is the point of data that is not geolocalized?

Geophysical maps, to be "georeferenced" in the long run, require someone with topographical skills. For metric precision, a "pocket" GPS (for walking) is not enough in absolute terms (just because it displays meters does not mean that its precision is metric). Topographers use differential-GPS systems that can relate a map to a reference base to the nearest centimeter.

Frequently, in relation to the objective, we only leave some reference posts whose lifespan ranges from some hours to a few months. Thus, we need to rely on someone skilled in topography in order to carry out geophysical prospecting on a long-term basis.

1.3.4. *Mapping types of prospecting: point-by-point, "walking" or drones*

If studies have led us to become accustomed to studying continuous functions, from negative infinity to positive infinity (for example, $f(x) = x^2$, defined for every x (therefore, an infinity) over the definition interval or domain $]-\infty, +\infty[$), our human nature requires us to limit our experimental expectations and only measure a parameter for a finite number of points. In every case without exception, geophysicists gather a finite number of measurements (besides, with a limited number of significant figures). Similarly, the number of parameters is finite, even if at the end this gives us the impression of continuity.

Therefore, prospecting always involves point-by-point measurements, but the vocabulary reserves this terminology for a process of elementary measurements, where one measurement is in some way manually taken after the other.

In a "point-by-point" approach, strictly speaking, whether in profile or in maps, we define an interval and a grid according to the resolution criteria (see the issue of the measuring grid mentioned above). At the hectare scale, we will use a pair of measuring tapes (a 20 m one, a 50 m one or a larger one, in practice!) laid out perpendicularly to the ground (with an optical square or the Pythagorean theorem) (Figure 1.8).

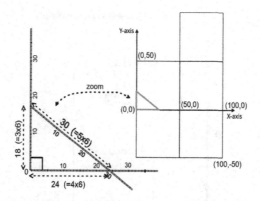

Figure 1.8. *Creating a right angle with the Pythagorean theorem and then using (50 m) measuring tapes allows us to divide a site into squares as small as a few hectares. For a color version of this figure, see www.iste.co.uk/florsch/geophysics1.zip*

The practical values are the multiples of (3, 4, 5). For every n integer other than zero, we obtain $3^2 + 4^2 = 5^2 \Leftrightarrow (3n)^2 + (4n)^2 = (5n)^2$. The figure shows an example where n = 6, thus using the triplet (18, 24, 30). It is more accurate to consider the largest possible triangle, and (30, 40, 50) represents the best choice for 50 m measuring tapes.

In so-called "walking" mode, the device and the procedure can sample rapidly, giving the impression of continuous measurements, but actually proceeding at a rate of 10 or 20 samples per second. In cartography, we create parallel and regularly spaced out profiles, and the procedure is set for measurements that are regularly spaced out. This can be done with a metronome incorporated to the device or even while listening to music. Individuals walk with steady steps over the distance, so that it is enough to give them the beat, like in a march, to obtain a constant speed and finally see to it that the measurements, sampled every Δt, are regularly spaced out during the march as the pace is constant.

Therefore, the procedure very often follows a technique called boustrophedon (#): U-turns like oxen tilling the fields. This technique is described in Figure 1.9.

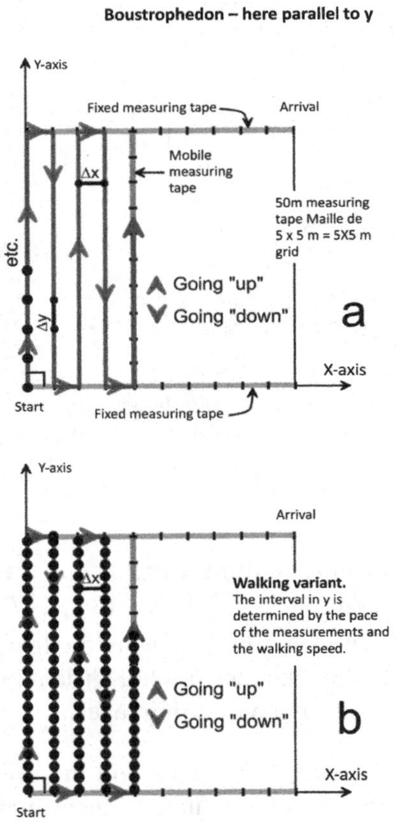

Boustrophedon – here parallel to y

Figure 1.9. *The boustrophedon movement is the most effective way of collecting data on a regular grid. For a color version of this figure, see www.iste.co.uk/florsch/geophysics1.zip*

Let us see how these rules have been respected in the first example (section 1.1.3).

First, the method: we know that the Roman brick walls are significantly more electrically resistive than the clay-marl enclosure of the site. The map clearly reveals this, showing conducting areas at less than $20\,\Omega m$ in blue and at more than $50\,\Omega\,\mathrm{m}$ in brown. This contrast is not significant in view of the resistivity found in nature. In fact, it turns out that the contrasts specific to the walls are already somewhat masked here by the cover of electrically homogeneous earth, which is on average 30 cm thick at this site. Apparent resistivity produces a sort of average of the resistivities under the electrodes (see above the definition of apparent resistivity).

Second, the depth of investigation. The depth of the walls ranges from 30 cm to maybe 1 m, and their upper side does not reach a depth of 50 cm. A device with a depth of investigation of 50 cm is suitable, and this is the case here with $C_1P_1 = 1$ m.

Third, the wavelengths of the anomalies. A characteristic width for a wall is around 50 cm. With a roof at a depth of 30 cm, we can typically expect a "width anomaly" in the range of 1 m.

Fourth, as meters are appropriate, there is no point in choosing a finer scale, except maybe for subzones where we would like to see more details. However, the essential aspect for archeologists here is the *detection* of structures more than the detailed creation of a map, which will nevertheless require excavations (rescue or programmed excavations, but always officially authorized).

If it is not necessary to touch the ground, drones can be used: this approach is clearly developing, for example in magnetism, and it can actually take measurements faster. The boustrophedon approach is used, with redundant data cross-checked against a certain number of profiles to correct potential temporal variations.

1.4. Adapting the models to the targets

Section 1.3 focuses mainly on the adaptation of scales. We also need a physical type of adaptation, which can be expressed in simple terms: the method chosen must accurately measure (be sensitive to) the parameters that vary from the target to the environment.

Thus, magnetic prospecting cannot be carried out to find water[12]. Even if we qualify this statement by pointing out that the magnetic approach provides information about geological structures, which are possibly suitable for aquifers, the source of the signal (and of the anomaly in general) is still not water. The adaptation will also be an issue related to the dimension of the source.

Let us imagine some small but highly magnetic waste, for example the iron residues of a smelter, scattered around in a soil. Even with a low proportion of these materials (1%) and a representative volume of 1 m^3 (as a reference point), the magnetic anomaly produced will be easily detected. With this dissemination, these particles, despite being extremely conducting compared to the enclosure, will be electrically invisible, as the current path deviations they will cause in relation to the part played by these current paths in the enclosure will be negligible. On the contrary, an aluminum frame a few meters large and with all its electrical connectivity (in one piece electrically speaking), can be detected with electrical − or, even better, electromagnetic − methods (Slingram, see Volume 2 (details in Bibliography)), whereas the magnetic response will be negligible.

All these remarks show that geophysics certainly cannot do everything. Let us recall here the "hypothesis elimination" aspect that represents simultaneously the strength and the limitations of geophysics. Thus, the absence of magnetic anomaly proves that a heap of waste contains no iron (but it does not demonstrate that there is no metal). Similarly, the lack of electrical anomaly in a magnetic area proves that no large, conducting body is buried here, or if it were, it would be entirely insulated.

---

12 We will not mention here the "science" of dowsers and other diviners, which is not statistically tested (despite what they occasionally think, which makes them somewhat recognized figures, and that explains it). However, some water diviners sometimes make an unformalized summary of "practical" and cumulative knowledge about nature.

Direct Current Electrical Methods

2.1. The electricity used in geophysics[1]

2.1.1. *The current flowing in the ground*

In every metal and some other materials, such as graphite or pyrite, mobile electrons ensure the conduction of electricity. At room temperature, they constitute an actual electron "gas". When a voltage – we also refer to potential difference – is applied to such a conductor, the voltage source is actually a source of electrons (of chemical origin in a battery), which will move slowly within the solid, a few millimeters per second, as they are slowed down by matter itself.

When we turn on the lights on the ceiling, we do not wait for these electrons, which have traveled from the switch, to reach the ceiling. This is due to the very rapid speed at which the electrons already in the wire are put into motion. It depends to a certain extent on the environment and it is generally a substantial fraction of the speed of light (for example 66% of c in a classic coaxial cable).

The theory applied to electricity is Maxwell's theory. Two links on Wikipedia may serve as an introduction for curious readers:

https://en.wikipedia.org/wiki/Electromagnetism for electromagnetism and

---

1 As a complement to this chapter and the following ones, the reader can, for example, refer to http://appliedgeophysics.berkeley.edu/dc/index.html and, for deeper notions in resistivity data, to https://pubs.usgs.gov/pp/0499/report.pdf.

https://en.wikipedia.org/wiki/Speed_of_electricity for the "speed of electricity".

In the subsoil, "electronic" conductivity is rare, since the materials that carry this type of electricity are rare. The only exceptions are some minerals in metallic deposits or specific polluted areas. Otherwise, the electric charge is nearly always carried by ions, dissolved in water, which can be found almost everywhere in the ground. Should water be totally absent, electric current will be badly conducted and will use other means, so that similar bodies will be regarded by surface geophysicists as highly resistant (unweathered limestone or granite, for example). Thus, although in metallic elements it is electrons that physically move (on the contrary, an "electron hole" will move in the opposite direction, but the latter does not correspond to a material particle – at least in classical physics), in electrolytes, namely in liquids (in most cases, water) where ions have been dissolved, cations (positively charged ions) and anions (negatively charged ions) moving in opposite directions ensure that current flows. The classic example involves NaCl. NaCl is solid in a dry state (salt) and even in a hydrated state (coarse salt crystallized in a hopper shape). In a solution, the strong polarity of the water molecules (see, for example, https://en.wikipedia.org/wiki/Chemical_polarity) dissociates and disperses NaCl into cations (Na^+) and anions (Cl^-). The latter are surrounded by a certain number of water molecules: the process of hydration. It is Coulomb's law that makes molecules and atoms stick together[2].

In terms of electrodes, therefore, the nature of the current changes: it is electrons that reach the negative electrode. They leave the electrode and combine with the cations present in the soil, such as sodium, according to the chemical reaction $Na^+ + e^- \rightarrow Na$. However, the resulting molecule of sodium reacts immediately with water, according to the reaction:

$$Na + H_2O \rightarrow (Na^+, OH^-) + \frac{1}{2}H_2^{\nearrow}.$$

Electrolysis (or a "Faradaic reaction") takes place. At the other electrode, it is Cl^- that yields its electron to the metal and becomes chlorine

---

2 Why is it not the NaCl molecule that dissociates water? Because the *ionic bond* that links the Na and Cl atoms is much less strong than the *covalent bond* that associates two hydrogen molecules with an oxygen molecule.

again. According to its metal, the electrode may be involved in electrolysis and either be corroded or become the site of a specific deposit. However, geophysicists are not interested in this phenomenon, which only affects the areas closest to the electrode. What matters is that there are two opposed ion currents, which constitute the electric current. The usual convention applies to this case: current moves from the positive electrode (cathode) to the negative electrode (anode); it is positive ions (cations) that travel toward the (–) electrode, whereas negative ions (anions) move toward the (+) electrode.

To find out more about the currents moving in electrolytes(#), several bibliographical sources can be found online.

Let us go back to our soils and rocks. As we can see, the higher the water content of the rock, the higher the ion content of the water and the more conducting the rock. These conductors must also be connected, as isolated water bubbles in a vitreous rock, for example, will not allow ions to be transported. A connected porosity filled with water represents a "volume" type of conductivity, as we can easily imagine (even if water is dispersed in the porosity of the rock), but there is also another type of "surface" conductivity, which is less apparent. It is a kind of conductivity that involves the surface of the component grains of the environment which, due to electrostatic effects linked to the "electric double layer"(#), attract more or less strongly the surrounding ions. The ions that are nearest the surface of the grain remain attached to the surface, but just a little bit farther (a dozen nm), there is a whole reservoir of electric charges (cations and anions) that remain mobile. Therefore, in a sandy soil with a slight clay content – which is, however, dry – there is a sort of connected network of surface conductors that allow current to cross the macroscopic rock, despite the absence of free water.

2.1.2. *Resistivity*

Let us recall here the notion of resistivity. It is the ability to resist the flow of an electric current. If we consider the classic example of a cylindrical piece of wire, based on a few basic principles, we can establish that the electrical resistance R of such a cylinder is proportional to its length and inversely proportional to its section (lateral surface). The proportionality coefficient is resistivity, which is in most cases called ρ (rho) and expressed in Ωm, so that we obtain:

$$R = \rho \frac{L}{S}.$$

Just like electrical conductance is by definition the inverse of resistance, conductivity, most often noted as σ (sigma), is the inverse of resistivity and is expressed in Siemens per meter (S/m).

Countless studies, attempts and models have been put forward to link conductivity (or resistivity, its inverse) to the water content of earth materials, as water (with its electrolytes) is the key factor for surface formations.

The authors will consider a practical formula which, despite not being universal, allows us to work with the right orders of magnitude. We show this relationship in terms of conductivity, as it is in relation to conductivity that volume conductivity is added to surface conductivity[3]:

$$\sigma_{rock} = \frac{\sigma_{water}}{F_1} S^n \Phi^m + \sigma_s.$$

In this equation, which sums up several works[4]:

– σ_{water} is the conductivity of the permeated water in the rock (which could be obtained by letting a rock sample drip), often known thanks to a well.

– F_1 is a constant called the "formation factor".

– S is the water saturation, ranging from 0 to 1. When S = 0, there is no water, whereas when S = 1, the rock is totally saturated. S is the proportion of water compared to the total saturation.

– Φ is the "effective porosity"(#)[5].

---

3 This is a rule for parallel conductance. It is acknowledged and tested that these types of surface and volume conductivity can be added to each other, but it is hard to provide a rigorous demonstration due to the complex arrangement of the grains and empty spaces (saturated with water) of a granular rock. For example, it is necessary to refer to the *effective medium theory*(#).

4 Rhoades *et al.* (1989), Frohlich and Parke (1989), as well as Mualem and Freidman (1991). See Bibliography for details.

– n and m are two empirical constants, which depend on the rock and other factors. They are often in the range of 2.

– σ_s is the surface conductivity.

Figure 2.1 shows a porous medium. We should be careful with this type of representation: a 2D stack of discs of the same radius does not seem to let anything pass, but a stack of identical spheres certainly allows grains to flow. Two-dimensional (2D) and three-dimensional (3D) properties may differ substantially for material or natural materials. Moreover, the 2D structure would either have no thickness (like an image) or be infinitely long perpendicularly to the table, making the density of matter entirely dependent on (x,y) (without z).

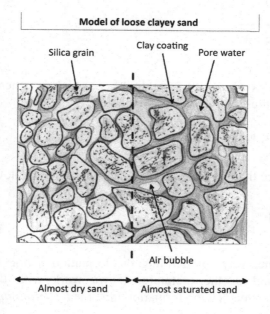

Figure 2.1. *In the absence of clay, current flows because of electrolytes (water + ions). The clay coating offers a special path for electricity to pass. If the soil is well drained, it is even the only contributing factor. For a color version of this figure, see www.iste.co.uk/florsch/geophysics1.zip*

5 A vitreous rock, enclosing water particles, would prevent ions from flowing, and would therefore be resistant. "Effective" porosity refers to a type of porosity, i.e. a set of interconnected pores.

Clay presents very high surface conductivity. The aforementioned formula is in line with the fact that a dry rock may be conductive, in relation to its clay content. Besides, studies such as those conducted by Rhoades *et al.* suggest that the clay content should be directly linked to surface conductivity, for example with the following formula:

$$C_a = \frac{\sigma_s + 2.1}{2.3},$$

where C_a is the clay content.

Besides, resistivity is naturally:

$$\rho_{rock} = \frac{1}{\dfrac{\sigma_{water}}{F_1} S^n \Phi^m + \sigma_s}.$$

When surface conductivity is negligible – which is generally the case when clay is not present – if $\sigma_s = 0$, we obtain:

$$\rho_{rock} = F_1 \rho_{water} S^{-n} \Phi^{-m}.$$

We can deduce that the factor F_1 may be calculated as follows:

$$F_1 = \frac{\rho_{rock}}{\rho_{water}} S^n \Phi^m.$$

There are other ways of defining the formation factor, and different definitions abound in the literature. If we work with the saturation hypothesis, we assume that the formation factor is *by definition* the relation:

$$F_2 = \frac{\rho_{rock}}{\rho_{water}}.$$

This is a landmark and empirical definition. Therefore, Archie's law means that: $F_2 = a\Phi^{-m}$. The parameter (a) is called "tortuosity".

However, despite being historically more rigorous (and deriving from petrophysics), this definition raises some problems when clay is present, which is the case more often than not. It is as if this definition were overused, as several works or articles employ it and/or rely exclusively on it, without always recalling that its applicability is limited to clay-free environments or without offering any alternative.

If we consider all of the studies to date, we cannot help but point out that there is no expression that is truly universally accepted, and that it is simply better to specify in our works and research which formula is being used.

Figure 2.2 shows some resistivity values found in nature. A resistivity of 100 Ωm is representative for calculations that involve orders of magnitude. Let us bear in mind that the resistivity of copper is 10 billion times lower.

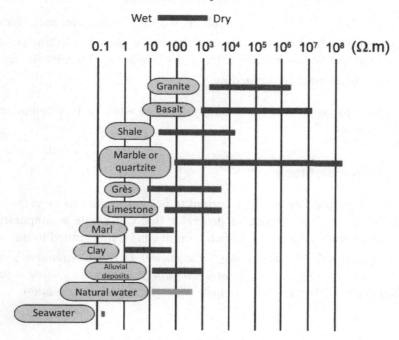

Figure 2.2. *Resistivity values commonly found in nature. For a color version of this figure, see www.iste.co.uk/florsch/geophysics1.zip*

2.1.3. *Separating volume conductivity from surface conductivity*

As a measurement provides a unique voltage resulting from a transmitted current, we cannot see *a priori* how the two types of surface conductivity can be separated[6]. However, there are some cases where this is possible. In very humid or rainy weather, the soil presents a degree of near saturation. Conductivity can be noted as:

$$\sigma_{rock} = \frac{\sigma_{water}}{F_1} S^n \Phi^m + \sigma_s = \sigma_V + \sigma_s.$$

When rain has not fallen in a long time, the water content is on the contrary very low, so that:

$$\sigma_V = \frac{\sigma_{water}}{F_1} S^n \Phi^m \cong 0,$$

leaving only $\sigma_{rock}^{dryweather} \cong \sigma_s$. This result, as we have seen, is an indicator of the clay content. The difference between the map obtained in a rainy period $(\sigma_V + \sigma_s)$ and the map obtained after a dry spell (σ_s) reveals the areas where water is mobile and drained.

To obtain these results, we naturally need to work in two completely different seasons.

2.2. Ground resistance

Electrical prospecting requires current to be sent to the soil by means of two electrodes. Let us consider a generator that can create a temporarily direct voltage expressed as V_0. Which current can be transmitted to the soil with this generator? *Is it dangerous? The answer is yes. Consequently, we will dedicate a specific paragraph to safety measures below, after introducing the notions required to take this aspect into consideration.*

---

6 This is one of the goals of the "Induced Polarization" method, which is still being researched.

2.2.1. *A first approach using hemispherical electrodes*

The basic outline of a transmission circuit connected to the ground was illustrated in Figure 1.4. However, it assumed the presence of point electrodes. In terms of electricity, the terminals of the generator terminate in an electrical circuit where we can consider three types of resistance one after the other: the two electrodes and the environment itself. Let us start by seeing what happens with only one electrode. To make the calculations involving the earth connection easier, we can assume that the return electrode is at an infinite distance and defines zero potential at this infinite distance itself (in practice, we could imagine that the cable reaches the sea, where it is connected to a huge grid at the bottom of the sea. We can guess that it will not be this electrode that will oppose strong resistance to the current).

The current that will flow is given by Ohm's law: $V_0 = R_T I$, where I is the current and R_T is by definition the resistance of the earth connection. Thus, we need to estimate the latter. This requires us to briefly carry out integration, as illustrated in the frame below in relation to a hemispherical electrode of radius (a) (see Figure 2.3).

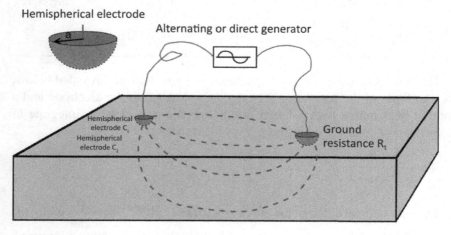

Figure 2.3. *A transmission system connected to the ground with non-point electrodes, i.e. realistic electrodes. Choosing hemispherical electrodes makes it easier to calculate the orders of magnitude of ground resistance. For a color version of this figure, see www.iste.co.uk/florsch/geophysics1.zip*

For a soil of homogeneous resistivity (at least in the immediate vicinity of the electrode), ground resistance is equal to:

$$R_T^{\text{hemisphere}} = \frac{\rho}{2\pi a}.$$

It naturally represents the resistance of the electrode with *all* the rest of the ground to infinity. In fact, it constitutes the resistance of the ground *outside* the highly conductive hemispherical boundary of the electrode.

Let us demonstrate this formula.

Let us consider a hemispherical electrode of radius (a) and infinite conductivity. The resistance of the electrode is equal to zero, and the electrode is in contact with the resistive soil. Let us trace an imaginary shell between r and r+dr centered on this electrode. The *resistance* of this shell is easy to estimate as this flat shell looks like a cylinder whose thickness is dr and whose surface is $S = 2\pi r^2$. Its resistance dR is given by the aforementioned formula applied to the length dr: $dR = \rho \dfrac{dr}{S} = \rho \dfrac{dr}{2\pi r^2}$, as this is naturally the surface of the hemisphere.

The resistance, from the electrode to infinity, is therefore:

$$R = \frac{\rho}{2\pi} \int_a^{+\infty} \frac{dr}{r^2} = \frac{\rho}{2\pi} \left[-\frac{1}{r} \right]_{r=a}^{\infty} = \frac{\rho}{2\pi a}.$$

Besides, because of a small variation in the calculations provided below, we can find out the resistance between the radius (a) of the electrode and a parallel hemisphere with a larger radius (b). We only need to integrate to (b) instead of considering infinity. We obtain that:

$$R_a^b = \frac{\rho}{2\pi} \left(\frac{1}{a} - \frac{1}{b} \right).$$

Let us compare then the total resistance $R_T^{\text{hemisphere}} = \dfrac{\rho}{2\pi a}$ with the resistance obtained when we stop at the distance (b). Let us choose, for example, b = 0.5 m for a = 0.05 (or more generally: b = 10a). We obtain that:

$$R_T^{\text{hemisphere}} = \frac{10\rho}{\pi},$$

whereas:

$$R_a^b = \frac{\rho}{2\pi}\left(\frac{1}{a} - \frac{1}{b}\right) = \frac{\rho}{2\pi}\left(\frac{1}{0.05} - \frac{1}{0.5}\right) = \frac{9\rho}{\pi}.$$

This resistance is 90% of the previous value, for which we considered infinity. This small experiment illustrates a general property: *ground resistance is virtually entirely confined to the close proximity of the electrode.*

2.2.2. Realistic electrodes: metal posts

We rarely use hemispherical electrodes, as we prefer to employ cylindrical metallic electrodes of diameter (d) and length $L \gg d$ which, once buried, offer a resistance in the range of:

$$R_T^{post} = \frac{\rho}{2\pi L}\left[Log_e \frac{8L}{d} - 1 \right].$$

This formula is a good approximation, and a good demonstration can be found in a book published by Dwight in 1931 (but unfortunately not very affordable)[7].

In practice, in a typical, small prospecting operation, we will sink a 50 cm-long metal post with a 1 cm-diameter 25 cm into the ground. The ground resistance is then:

$$R_T = 2.73\rho, \text{ or } 273\Omega \text{ for a ground resistivity value of } \rho = 100\Omega \text{ m.}$$

We need two electrodes, so that the resistance of the electric circuit, everything else being equal, will be double.

These values help us calculate orders of magnitude, but the reality is less simple. When we sink a metal post into the ground, which is in most cases somewhat granular, some parts of its surface will certainly touch the ground, whereas others will not. Quite systematically, this theoretical value will have to be (approximately) doubled. We cannot identify with certainty

---

7 However, a suitable document can be found at: http://www.bu.edu.eg/portal/uploads/Engineering,%20Benha/Electrical%20Engineering%20/2369/publications/Tamer%20Mohamed%20Elsaid_1.PDF.

which parts of the surface of the electrode originate the current, unless there is good contact everywhere (which is the case for clay).

2.3. The basic array for electrical prospecting

When we connect the generator to the two ground resistance electrodes R_T, a current will flow. It will be in the range of:

$$I = \frac{V_0}{2R_T} = \frac{V_0}{R_{TOTAL}},$$

if we consider a homogeneous soil, equal ground resistance values, etc. However, despite depending naturally on the resistivity of the soil, ground resistance values also depend on how the electrodes have been sunk, their diameter and the actual contact with the soil. The current measurement I is representative of all these parameters, but this is irrelevant if we want to find out the resistivity of the subsoil[8]. This also means that we cannot use a multimeter in ohmmeter mode to measure the resistivity of the soil (moreover, it would be saturated by the polarization of the electrodes...do try!).

To find out the resistivity of the subsoil (which we will assume to be still homogeneous for a little longer), we need to use four electrodes in total, allowing us to avoid ground resistance.

We need to take several steps to establish this protocol.

2.3.1. *Stage 1: what is the potential at the distance (r) from the electrode?*

Let us work with a hemispherical electrode, which is practical because of its symmetry, taken to the potential V_0.

An electrode could not have the shape of a point, as ground resistance would become infinite and therefore current would no longer flow... However, we use this notion, so we acknowledge that we transmit a current I to this electrode, which always presents its own resistance R_T. We merely look at the electrode from a distance, so that it looks tiny (more precisely,

---

8 This means, due to other factors as well, that we can in no way use a multimeter in ohmmeter function to measure anything interesting about the subsoil.

the electrode is small in relation to the scale of the ground and the inter-electrode distance).

After taking these precautions, we can admit then that we transmit a current by means of a point electrode. This is only a "practical" theoretical line of thinking.

Let us introduce *current density*, which is nothing more than the current per surface unit, and let us point out the similarities between this problem and the point electrode. The total current is I. It spreads with a current density J (in ampere per square meter) over the hemisphere centered on the electrode. J is symmetrically the same all over this hemisphere. The current flowing from the electrode must necessarily cross the surface of this hemisphere, which corresponds to $2\pi r^2$. Consequently, the current density is:

$J = \dfrac{I}{2\pi r^2}$. This significant result proves that current density decreases in relation to the square of the distance from the electrode.

Reasoning always in terms of symmetry, the potential only depends on r. Here, we will only present some simple calculations.

In a small cone whose tip is on the electrode, whose length is δr and whose small surface δS is an element of the sphere, the potential difference δV is given by Ohm's law: $\delta V = -\delta R\, \delta I$, where δR is the resistance of this cylinder and where the current flowing through the small surface is δI. The minus sign is necessary as the potential decreases as we get farther from the electrode. The infinitesimal resistance δR is equal to:

$\delta R = \rho \dfrac{\delta r}{\delta S}$.

Ohm's law can then be written as: $\delta V = -\delta R\, \delta I = -\rho \dfrac{\delta r}{\delta S} J \delta S = -\rho J \delta r \Leftrightarrow -\dfrac{\delta V}{\delta r} = \rho J$.

Moving on to the derivatives by passage to the limit of the quotient quantities, we obtain here:

$-\dfrac{dV}{dr} = \rho J$, or: $J = -\dfrac{1}{\rho}\dfrac{dV}{dr} = -\sigma \dfrac{dV}{dr}$ [9].

---

[9] This is the specific formulation of Ohm's law written in continuous mediums in the vector form $\vec{J} = \sigma \vec{E}$ with $\vec{E} = -\vec{\nabla} V$, where, in the general case, σ is a tensor.

Replacing J with its value, we obtain:

$$\frac{dV}{dr} = -\frac{\rho I}{2\pi r^2} = \frac{\rho I}{2\pi}\left(-\frac{1}{r^2}\right)$$

Now, the function whose derivative is $-\frac{1}{r^2}$ is simply $\frac{1}{r}$. We can deduce that the

appropriate potential is: $V = \rho\frac{I}{2\pi r} + C$, where C is an integration constant. The

convention is that the potential becomes zero as we approach infinity, so that

$C = 0$ (be careful with the potential, which is first of all a mathematical notion)[10].

Let us consider this important equation: the potential created by a point electrode in a homogeneous (and isotropic) medium of resistivity ρ, at a distance r from the electrode, is equal to:

$$V(r) = \frac{\rho I}{2\pi r}.$$

This represents the end of stage 1.

2.3.2. *Stage 2: describing "point" electrodes in more detail*

The aforementioned equation only applies outside the hemispherical electrode. Within, the potential is equal to V_0, the tension of the generator, and all of the electrode (almost infinitely conductive, given the resistivity contrast between copper and the soil) has the same potential. Therefore, there is *never any reason* to assume that (r) may approach 0. The equation that we have already obtained for a hemispherical electrode:

$$R_T^{hemisphere} = \frac{\rho}{2\pi a}$$

---

10 In any case, we cannot measure an absolute potential but only a potential difference. We can see here that the calculations involve integration – the opposite operation of derivation – and that the initial equation already focuses on the potential difference dV. Moreover, our instruments, i.e. voltmeters, are actually ammeters that measure very weak currents. In order for us to take a measurement, electrons *must* flow. Finally, the potential is always defined in relation to a reference point, an integration constant, so that there is an infinite number of possible potentials but only one *potential difference*, which is measured thanks to a current.

shows that the potential is constant up to r = a, and that it then decreases following $V(r) = \dfrac{\rho I}{2\pi r}$ for r > a, as illustrated in Figure 2.4. This figure was drawn from an excellent document (in French) called *"Principe de conception et de realisation des mises à la terre"* ("Design and implementation principles of grounding systems")[11], which can be easily found online. A valuable reference in English would be the book by G. F. Tagg, 1964[12]. Another synthesis can be found here: http://dl.lib.mrt.ac.lk/bitstream/handle/123/776/93877_5.pdf?sequence=7.html.

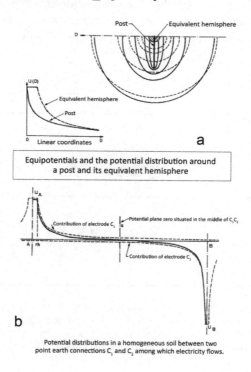

Equipotentials and the potential distribution around a post and its equivalent hemisphere

Potential distributions in a homogeneous soil between two point earth connections C_1 and C_2 among which electricity flows.

Figure 2.4. *Patterns of the potential with hemispherical electrodes or posts. The potential follows 1/r exactly outside a hemispherical electrode and at a certain distance from a post electrode. a) Geometry of the equipotentials near the electrode; b) graph of the potential for two transmitting electrodes (+I) and (–I). We should recall that most drops in potential take place near the electrodes*

---

11 http://chris.murray.free.fr/H115.PDF
12 G.F. Tagg, 1964. *Earth Resistances*. Georges Newnes Ltd, London. Also published by Pitman Publishing Corporation, the same year, pp. 258.

However, geophysicists usually employ the potential formula in 1/r, as they are well outside the electrode. If we consider a hemispherical electrode in a homogeneous environment (at least near the electrode), the equation remains valid outside the electrode, regardless of its radius. However, we should point out that if this radius is too small, the resistance increases and, for the same generator voltage, there is gradually less current. This leads us to increase the diameter of the electrode. It is common practice to dig a little hole, typically with a diameter of 10–20 cm, filled with water-saturated bentonite(#)[13], with the copper or inox electrode placed in the middle. The resistivity contrast between the bentonite and the enclosure is such that we can virtually regard the electrode as a hemisphere, especially if we also remembered to add some salt to the clay (Figure 2.5).

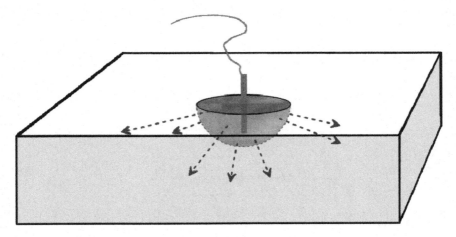

Figure 2.5. *Digging the ground in a bowl shape (with a diameter ranging from 5 to 20 cm), filling it with bentonite (wet and possibly slightly salted) and planting an electrode made of copper or another metal is a very effective method to make very good contact with the ground*

Finally, the question of the "point" shape of the "post" electrode refers to another situation, since the electrode is no longer hemispherical, regardless of the homogeneous nature of the material of the enclosure. In

13 Most cat litter trays are made of bentonite: this is a practical source for geophysicists. However, bentonite can also be found in bags, like cement.

practice, this electrode looks like a point when "seen from afar", so that for $r \gg L$ (let us say by a factor of 10), the equation $V(r) = \dfrac{\rho I}{2\pi r}$ remains valid. This idea is well developed in Figure 2.4, as we have already seen.

When we plant some post electrodes in a row, as is the case for panel arrays (mentioned further on), we usually recall that the electrodes must be spaced out five to 10 times their length, and the more the better. However, we can see the problematic compromise that this entails: this process takes place in the right conditions if we sink the electrodes less deep, to the detriment of ground resistance, which increases.

2.3.3. *Stage 3: setting up all the electrodes: two transmitting electrodes and two measuring electrodes*

The electrical circuit must be closed for the transmission, so we need two electrodes: one is conventionally positive (noted as C_1) and another, called the "current-feedback" electrode, is negative and noted as C_2. The return electrode C_2 is the same as the electrode C_1, but here the current flows from the ground rather than into the ground. Therefore, the potential generated by the electrode C_2 is negative.

As we can only measure *potential differences*, we will use P_1 and P_2[14]. P_1 is conventionally attached to the (+) terminal of the voltmeter and P_2 to the (–) terminal. Thus, the latter is the point of reference used from an electrical point of view.

We have now four electrodes (C_1, C_2, P_1, P_2). Figure 2.6 illustrates how they are set up for carrying out mesoscale mapping. There must be more than one person to manage the four electrodes.

We generally assume that the voltmeter has very high impedance or, in other words, that plugging it does not modify the potential difference to be measured[15].

---

14 "C" stands for "Current" and "P" for "Potential".
15 This was not the case before the invention of digital multimeters. Therefore, "nulling method" had to be used to measure potential differences.

Modified from C. Meyer de Stadelhofen

Figure 2.6. *Mapping or sounding on a plurimetric scale in order to explore the first meters of subsoil. For a color version of this figure, see www.iste.co.uk/florsch/geophysics1.zip*

Figure 2.7 briefly illustrates a type of safety electrode that does not hurt our back. This tool comes in handy when we employ medium-size arrays (ranging from some decimeters to a few dozen meters).

How can the data obtained be exploited? Let us consider for now a homogeneous environment. The calculations that yielded $V(r) = \dfrac{\rho I}{2\pi r}$ remain valid if the point is on the surface. In other words, for a current I transmitted by the electrode C_1 and for a potential considered on the electrode P_1, we will write $V_{C_1 \to P_1}$ and obtain (the distance between C_1 and P_1 being $C_1 P_1$):

$$V_{C_1 \to P_1} = \frac{\rho I}{2\pi C_1 P_1}.$$

Similarly:

$$V_{C_2 \to P_1} = \frac{-\rho I}{2\pi C_2 P_1}.$$

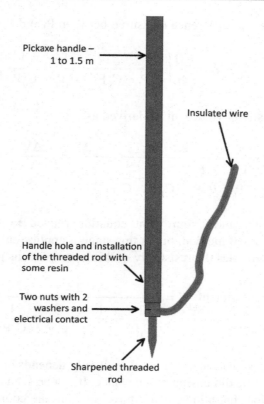

Pickaxe handle –
1 to 1.5 m

Insulated wire

Handle hole and installation
of the threaded rod with
some resin

Two nuts with 2
washers and
electrical contact

Sharpened threaded
rod

Figure 2.7. *Geophysical campaigns often involve thousands of measurements. This type of electrode (which may also be slightly sunk into the ground with a hammer) is safer and more effective*

The potential obtained for P_1 by transmitting and the extracting is therefore equal to:

$$V_{P_1} = V_{C_1 \to P_1} + V_{C_2 \to P_1} = \frac{\rho\,I}{2\pi}\left(\frac{1}{C_1 P_1} - \frac{1}{C_2 P_2}\right).$$

Similarly, for the electrode P_2, we obtain:

$$V_{P_2} = V_{C_1 \to P_2} + V_{C_2 \to P_2} = \frac{\rho\,I}{2\pi}\left(\frac{1}{C_1 P_2} - \frac{1}{C_2 P_2}\right).$$

The potential difference measured between P_1 and P_2 is then:

$$\Delta V = V_{P_1} - V_{P_2} = \frac{\rho I}{2\pi}\left(\frac{1}{C_1P_1} - \frac{1}{C_2P_1} - \frac{1}{C_1P_2} + \frac{1}{C_2P_2}\right).$$

The resistivity value can be derived as:

$$\rho = \frac{2\pi}{\left(\dfrac{1}{C_1P_1} - \dfrac{1}{C_2P_1} - \dfrac{1}{C_1P_2} + \dfrac{1}{C_2P_2}\right)}\frac{\Delta V}{I} = K\frac{\Delta V}{I}.$$

This is the most significant equation related to the so-called direct current electrical method, but at this stage it applies only to a homogeneous ground. It provides the resistivity of this homogeneous ground.

This coefficient:
$$K = \frac{2\pi}{\left(\dfrac{1}{C_1P_1} - \dfrac{1}{C_2P_1} - \dfrac{1}{C_1P_2} + \dfrac{1}{C_2P_2}\right)}$$

is called the *geometric coefficient*, as it only depends on the geometry of the array. It has the dimension of a distance. It can be seen as a sort of electrical normalization that allows us to move on from the couple of measurements $(\Delta V, I)$ to the resistivity of the ground by taking the geometry into account.

Let us point out that there are cases where K may be infinite and therefore ρ cannot be defined. For example, this happens if the electrodes P_1P_2 are situated on the bisector of C_1C_2.

We can see here that finding the resistivity does not involve any notion of electrode ground resistance. This is the goal. However, this notion is naturally involved, as it is the one that limits the current, but it leaves the ratio U/I unchanged.

This formula also allows us to *predict* the order of magnitude of the potential that will be measured in an experiment, as it is simply:

$$\Delta V = \frac{\rho I}{K}.$$

Assuming always that the ground resistances are equal to R_T, then the current $I = \dfrac{V_0}{2R_T}$ entails the potential $\Delta V = \dfrac{\rho V_0}{2R_T K}$.

Let us evaluate K in a specific case (the most commonly used types of geometry are described below): when the electrodes are aligned and spaced by a distance (a) in the order $C_1 P_1 P_2 C_2$ (called "the Wenner alpha array"). This arrangement resembles the one illustrated in Figure 2.6, which would require this exact arrangement $C_1 P_1 = P_1 P_2 = P_2 C_2 = a$.

Therefore, we obtain:

$$ K = \frac{2\pi}{\left(\dfrac{1}{a} - \dfrac{1}{2a} - \dfrac{1}{2a} + \dfrac{1}{a} \right)} = 2\pi a. $$

Thus, K increases with the size of the array: this is a general rule. For (a = 5m), K is equal to 31.4 m.

Finally, let us make a prediction for this example. With the following values: $\{\rho = 100\ \Omega m; V_0 = 12\,V; R_T = 500\ \Omega; a = 5\ m\}$, we obtain:

$$ \Delta V = \frac{\rho V_0}{2R_T K} = \frac{100 \cdot 12}{2 \cdot 500 \cdot 31.4} = 38\ mV . $$

This is a small voltage. In order to reach the percent accuracy, a digital multimeter must be able to display the tenths of millivolts, or the signal will have to be amplified.

In practice, resistivity values can vary widely. K is often much greater, leading to an even weaker signal to be measured. Let us recall then that *the range of potential differences to be measured goes typically from a fraction of millivolts to a few volts.*

This is when things get a bit complicated, as the potential obtained – started with our transmission – will not be the only one. Hence, stage 4!

2.3.4. *Stage 4: guarding against intruders*

2.3.4.1. *Spontaneous polarization and low-frequency stray currents*

Studied in another chapter as a geophysical method in its own right, spontaneous polarization (SP) is added to our own measurement and represents a significant disturbance, as this potential difference may be of the same order of magnitude as the signal produced on P_1P_2 and coming from the generator.

Let us approach SP in experimental terms. We only need to plant two electrodes in the ground, for example a few meters from each other, and link them to the high-impedance voltmeter. We can detect a signal that may range from a few millivolts to sometimes even 1 V. As we will see in Chapter 3, this potential difference may be caused by different factors:

– the "redox" potential deriving from electrode-electrolyte interactions in the ground. The whole system constitutes a battery. If the metals of the electrodes are different, the potential is high (for example, a half volt), but the differences in the nature of the ground electrolytes or in temperature at each electrode are enough to generate small potentials related to the Nernst equation(#) (a few millivolts);

– there is an electrofiltration potential[16], linked to the water flowing in the subsoil and in roots and plants (see Chapter 3, which is dedicated to SP). It may reach 1 V with root pumping;

– there is a potential generated in the subsoil by the variations in volume of the electrolyte concentration and by some minerals that form "geo-bacteria". It reaches a few millivolts;

– finally, there is a time-varying potential difference associated with *stray currents* and the *magnetotelluric* field(#), but it is rather weak and generally negligible if the distance between P_1 and P_2 is less than 100 m.

To take a good electrical measurement, we need to get rid of the effect of SP.

A first method involves, while we are taking an electrical measurement, measuring SP before supplying the current and then subtracting it from the

---

16 This is a common misconception: there is a potential *difference*, since this is what is measured and not the potential, which is in any case defined up to an additive constant.

measurement taken as the direct current was being supplied. It may be opposite in sign to the measurement, in which case we subtract a negative value, therefore adding in this case a positive value.

Another method involves making electricity flow in one direction and then in the opposite direction, and afterwards calculating the difference between the two measurements. The third method involves opposing a voltage to SP with a bridge circuit, so that SP can be compensated and the signal can be brought back to 0. Finally, the fourth method involves using a low-frequency alternating current. Measuring the resulting tension with an alternating voltmeter filters *ipso facto* the direct component, and therefore SP. This is a clever solution, and some diagrams provided at the end of the chapter are related to it. However, there are stray currents at all frequencies, especially near cities or power plants, which remain an inconvenience.

Before elaborating on the use of electrical methods, now that we know the orders of magnitude, it is necessary to talk about the dangers of electrical prospecting and the safety regulations and rules to be followed.

2.4. Dangers involved in electrical prospecting and safety regulations and rules

If necessary, find out about the notions of "phase, neutral and ground" in household installations. At home, safety measures protect us quite well against electrocution. We have double-insulated cables and there are earth connections on all devices whose frame is partly or completely made up of metal. Let us imagine a mains leakage in the frame of our washing machine. As the frame is on the ground, some current will flow from the phase to the ground, so that there will not be the same current flowing between the phase, where "the current comes from", and the neutral line, where "the current returns". If the difference is greater than 30 mA, the *differential* circuit breaker cuts the power, as it is supposed to do. If it does not work and we touch the frame of the washing machine, some current may flow through our bodies toward the ground or another object and electrocute us.

In the field, we are *on* the ground and send current *into* the ground. Consequently, *the safety tools available at home cannot be implemented in the field*.

At any moment, we risk becoming an electrode, inadvertently placing ourselves between the generator and the ground, and being traversed by electrical currents.

We need to protect ourselves from these risks and we cannot rely on safety regulations (or on the other operators present!). Here is a list of precautions and rules to follow:

– *Precautions*:

- Avoid working when the weather is too wet, as water facilitates the contact between our body, the ground and all the leakage currents.

- Whenever possible, do not exceed 10 mA currents. "Electric shocks" with this type of current are quite uncomfortable, but still not very dangerous (except for people with heart conditions, in which case everything depends on chance).

- Pay close attention to the transmitter if you are not near it. A commonsensical but simple rule is, when an operator is not near a transmitter, it means that we are not transmitting. Provided that this is one of those transmitters that cannot transmit if there is no operator there! When an operator is close to the transmitter, we should move away from the electrodes, so that we are not tempted to touch them.

- Be in the position to communicate with the person in charge of the transmission. Under 20 m, we can still communicate by voice, but as the distance increases, especially if we cannot see each other, walkie-talkies (like PMR 466) are required.

– *Rules:*

- Do not work if it is raining, as the current flows easily across wet surfaces! We cannot protect ourselves properly in rainy weather.

- Wear insulated shoes, such as rubber boots or ankle boots, which must be *leak-proof*. We need to remain insulated from the ground.

- Wear insulated gloves and clothes that cover your skin. This may be difficult when the weather is very hot.

- *Never* touch two points at the same time, for example with your left and right hand.

We should also keep well in mind that the electrode potential is the same as the generator potential – i.e. high – so that if we put one foot near the electrode and the other a little farther, and our shoes are not perfectly insulated, we once again risk being electrocuted.

Special precautions must be taken when the transmitting electrodes are far from the operator. There is always the risk that animals or people may have touched an electrode or removed a cable out of curiosity or by accident. We need to work based on what we can see, be well able to communicate and always be in a position to cut the power.

Naturally, there are regulations for commercial devices, such as a "panic button" that cuts all power. This is certainly a good thing, but it is also somewhat of an illusion, as once again we can only protect ourselves with our own means, because we physically touch the ground. The reality is that NO electrical device, whether commercial or built by ourselves, could meet safety regulations that protect us. Unless the measuring procedure is completely automated and no longer involves an operator.

If we respect all these rules, our safety is significantly improved, but there is always a certain component of risk. In the field, only we can ensure our protection from electricity. Finally, we should find out how to succor someone who has been electrocuted.

However, there is a fairly safe type of protection: use only voltages weaker than 20 or 30 V. The current will be generally weak (but unable to go very deep).

Finally, (nearly) every Ministry of Agriculture disseminates information pamphlets, which should be used by farmers. Besides the fact that animals (generally!) walk barefoot, they rarely read the safety directions. This is why these pamphlets, which are especially instructive, exist (and should be read by the owners). Here is a sample pamphlet: http://mrec.org/files/2011/02/farmsafe.pdf.

2.5. Apparent resistivity

We have considered the equation $\rho = K \dfrac{\Delta V}{I}$ which allows us, with four electrodes, to measure the resistivity of the soil, provided that it is homogeneous, by disregarding the effect of ground resistances. However, the subsurface is not homogeneous and, if it were, geophysics would be useless.

To move forward, we should represent the *current lines* as well as the equipotentials in a homogeneous subsoil. The current lines represent the paths of the charges moving from one electrode to the other. We can draw from Conrad Schlumberger's[17] landmark book his representation of current and equipotential lines, as illustrated in Figure 2.8.

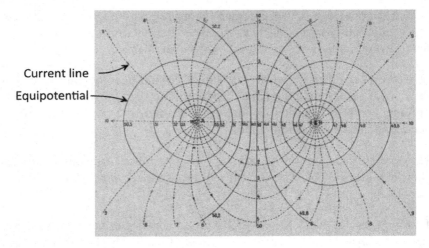

Current line

Equipotential

Figure 2.8. *Conrad Schlumberger's presentation of electrical quantities in a homogeneous subsoil. The solid lines represent the equipotentials, while the dotted lines represent the current lines. The + terminal of the generator is in A at 100 V, whereas the B terminal is set at 0 V. Here, the ground is seen from above, but if we only considered the lower part of the diagram for a vertical sectional view with the surface of the Earth, the geometry of the equipotentials and the current lines would be exactly the same*

17 We should now study the historical background: https://www.annales.org/archives/x/schlum.html. A translation in English is available, but without figures, at: https://archive.org/details/studyofundergrou00schlrich.

Even without any detailed calculations, we can easily see that the area explored by an array such as the Wenner array (four electrodes $C_1P_1P_2C_2$ in line and equidistant from (a)) only involves the immediate vicinity of said array, and the immediate surroundings are directly *under* the array.

For example, on both sides of a fault, at a given distance (let us say several times C_1C_2), the apparent resistivity will become more similar to the underlying resistivity.

However, things become more complicated when the array passes over a fault. Figure 2.9 shows a sample profile as we cross a fault that separates two soils. The apparent resistivity presents angulations that are not expected at first glance. On the left, the apparent resistivity noticeably moves toward that of the left part. On the right, the apparent resistivity will move toward the right. The currents are confined to one of the two areas. In the middle, the situation is intermediate, and the result is not as simple as we may hope when we imagine a smooth transition. However, let us point out that, if there were a (conductive or resistant) cover, these angles would be less sharp.

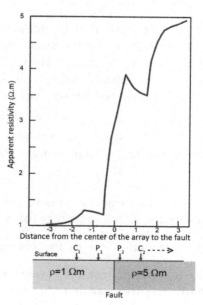

Figure 2.9. *Crossing a fault, the arrays do not provide generally smooth responses. This is due to the fact that the arrays themselves are made up of point electrodes, which suddenly change the environment as we follow the profile*

Calculations related to this type of response can be found in the "bible" of Van Nostrand and Cook's[18] landmark work.

Figure 1.4 illustrates a horizontally layered soil. When the array is small compared to the thickness of the cover, it can only "see", as it were, this superficial layer. On the other hand, when C_1C_2 is large, the currents penetrate almost directly all the way through to the underlying layer. The superficial layer plays a minor role in the potential measured, as it is crossed, but it is the deep layers that are determining.

In all these cases, we can see that the idea is to define an *apparent resistivity*. This merely involves the use of the formula $\rho = K\dfrac{\Delta V}{I}$, where we qualify that it is not the resistivity of a "small a" as "apparent":

$$\rho_a = \frac{2\pi}{\left(\dfrac{1}{C_1P_1} - \dfrac{1}{C_2P_1} - \dfrac{1}{C_1P_2} + \dfrac{1}{C_2P_2}\right)} \frac{\Delta V}{I} = K\frac{\Delta V}{I}.$$

When the equation is expressed in these terms, we can take into consideration the fact that resistivity is not constant in the ground. Some see apparent resistivity as a kind of "average resistivity" of the soils under the array. This approach is conceptually wrong and dangerous. If, when dealing – for example – with two soils whose respective resistivity is ρ_1 and ρ_2, we frequently obtain $\rho_1 < \rho_a < \rho_2$, this is not a general rule. It is actually quite easy to illustrate situations and arrays where apparent resistivity may be negative. Figure 2.10 shows one of these cases (which is unlikely but possible).

We may consider the following definition, which is sometimes provided: "apparent resistivity is the resistivity of a homogeneous soil which would yield the same ratio $\dfrac{\Delta V}{I}$ as the one actually measured". Unfortunately, a negative apparent resistivity value means that there is homogeneous soil with true negative resistivity.

---

18 https://pubs.usgs.gov/pp/0499/report.pdf.

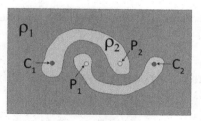

$\rho_2 \ll \rho_1$ $C_1P_1P_2C_2$ Wenner array

Figure 2.10. *The possibility of negative apparent resistivity values. The view is from above and presents two very conductive channels (filled with clay at 5 Ωm) dug in granite at 20,000 $\Omega \cdot m$. In a homogeneous or horizontally layered environment, with +I at the electrode C_1 and –I at the electrode C_2, the potential of P_1 is generally higher than that of P_2 due to the fact that it is closer to the positive electrode C_1, while the opposite is true for P_2, which is closer to C_2. Thus, we obtain $V(P_1) > V(P_2)$. Here, the presence of very conductive channels reverses this situation, so that $V(P_1) - V(P_2)$ may be negative. The apparent resistivity may then be negative. For a color version of this figure, see www.iste.co.uk/florsch/geophysics1.zip*

The only *good* definition of apparent resistivity was given by G. Kunetz in 1996 as the "ratio between the measured value of the parameter in question and its theoretical value in a homogeneous medium of unit resistivity".

In other words, for us it means that *apparent resistivity is* $\rho_a = K \dfrac{\Delta V}{I}$.

2.5.1. *A note about reciprocity*

Mathematicians demonstrate that if we swap the transmission pair (C_1, C_2) and the receiving pair (P_1, P_2), the apparent resistivity will be exactly the same, regardless of the array and the structure of the subsoil (relief included).

This property is used to confirm the measurements. We reverse C_1C_2 and P_1P_2 ($C_1 \Leftrightarrow P_1$ and $C_2 \Leftrightarrow P_2$), which means swapping the transmitter and the voltmeter. The current I changes (the ground resistances in P_1 and P_2 are different from those in C_1 and C_2), but ρ_a remains the same.

If, during this check, this is not the case, there may be a problem with the equipment or the cable, the SP may have been inadequately corrected, there may be an external parasite or we may be reaching the limits of the equipment.

2.6. Arrays in electrical prospecting: depth of investigation and sensitivity

The main arrays are represented in Figure 2.11.

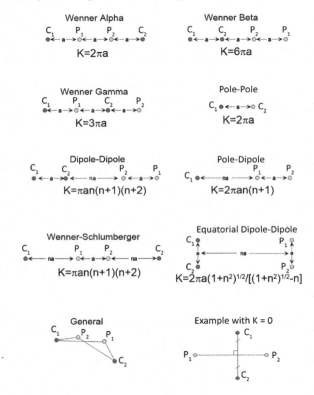

Figure 2.11. *Main electrical arrays. We should point out that when P_1P_2 is situated on an equipotential of C_1C_2, the potential difference is necessarily zero in homogeneous soil. The notion of apparent resistivity loses significance, as this result does not depend on the resistivity of the ground. The geometric coefficient is infinite. Despite everything, we can "work" with U/I ratios. For a color version of this figure, see www.iste.co.uk/florsch/geophysics1.zip*

Some arrays are more sensitive than others to artifacts that may mislead novices. Some instances are grouped in Figure 2.12. It is impossible to list every specific case. Ideally, to limit these potential, misleading effects, we should process all these measurements, profiles and maps by inversion, but this project is often too ambitious.

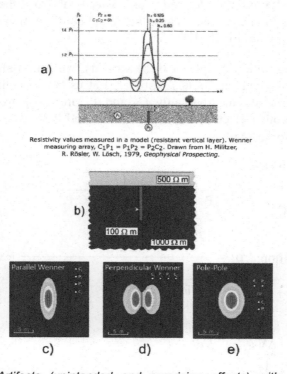

Figure 2.12. *Artifacts (unintended and surprising effects) with some arrays. a) Considered by Meyer de Stadeholfen, this example shows that the resistivity observed is not always intermediate among the resistivities found in the field. This example recalls, among other things, that it is dangerous to consider resistivity as a mean of the resistivities near an array. b) A model of conductive karstic fissure for cases (c)–(e). The fissure is 4 m long, 2 m deep and 5 cm thick. For convenience, the color scales are standardized. The contrasts obtained in (c)–(e) are slightly negative (–2.5% compared with the rest of the maps, represented in red). Consequently, these structures can hardly be detected, due to the presence of "geological noise". c) For the Wenner array, perpendicularly to the structure, the image is quite satisfactory. However, the direction of the structure is initially unknown. d) How can this rebound map be understood in relation to the perpendicular Wenner array? This is one of the main drawbacks of the Wenner array which, however, remains commonly used (we can also notice some rebounds in (a)). e) The pole–pole array yields the same map, regardless of whether C_1P_1 is parallel or perpendicular to the wall. On the other hand, the response contributes slightly less to the solution. For a color version of this figure, see www.iste.co.uk/florsch/geophysics1.zip*

We have emphasized that we need to use four electrodes: two transmitting electrodes (C_1 and C_2 transmission) and two measuring electrodes (P_1 and P_2). However, there are arrays that use two or three electrodes. The "missing" electrodes are placed far from the electrodes, which are close to one another. Thus, the actual configuration of the pole–pole array is shown in Figure 1.1.

To understand the principle of these variants, we need only reconsider the formula that yields apparent resistivity. Let us consider the pole–pole for example. We assume that only C_1 and P_1 are close to each other, whereas all the other electrodes are far from C_1 and P_1 as well as from one another. Therefore, we obtain:

$$\left.\begin{cases} C_1P_1 \ll C_1P_2 \\ C_1P_1 \ll C_2P_2 \\ C_1P_1 \ll C_1C_2 \end{cases}\right\} \Rightarrow \left| -\frac{1}{C_2P_1} - \frac{1}{C_1P_2} + \frac{1}{C_2P_2} \right| \ll \frac{1}{C_1P_1}.$$

In the equation: $.\rho_a = \dfrac{2\pi}{\left(\dfrac{1}{C_1P_1} - \dfrac{1}{C_2P_1} - \dfrac{1}{C_1P_2} + \dfrac{1}{C_2P_2} \right)} \dfrac{\Delta V}{I}$, we are only

left with:

$$\rho_a = \frac{2\pi}{\left(\dfrac{1}{C_1P_1} \right)} \frac{\Delta V}{I} = 2\pi a \frac{\Delta V}{I},$$

where we established that $C_1P_1 = a$. This also yields the geometric coefficient of the pole–pole array with $K = 2\pi a$.

2.6.1. *Depth of investigation and sensitivity*

To grasp the depths of investigation of the different arrays, we need to take into consideration simultaneously the transmitting and the receiving electrodes, and especially their relative positions. Often, it is enough to analyze the configuration of the currents by taking into account the position of the receptors in relation to these currents. At the same time, this raises the question of the spatial sensitivity of the array, which is in a sense the lateral counterpart of the depth of investigation.

These questions have been at the center of several papers in the scientific literature, such as Dahlin and Bing's[19] document (2003), which can be found online. Loke's tutorial is very thorough, but harder to read[20]. Finally, we should mention a poster written by R. Mota that prevents us from harboring any illusions[21].

There are two classic ways of defining the qualities of an array.

The first, which may be called the *sensitivity section*, involves quantifying the influence of a small volume (compared with the interelectrode distances) situated under an array and variable resistivity. Figure 2.13 provides an example of these sections for Wenner arrays. These sections can be obtained by charting the apparent resistivity in relation to the influence and position of this small volume. The red areas are those where an increase in the resistivity of the small volume increases the apparent resistivity, whereas the opposite is true for the blue areas.

If studied carefully, this type of section allows us to recognize some artifacts, such as the paradoxical "rebounds" (also called "ghosts") that can be observed with this array. These diagrams are mostly counterintuitive, contributing to their significance, but also simultaneously hard to use.

The second approach involves studying the response of an array in relation to a confined volume whose size, however, is significant for the array used: we are considering a realistic situation. Yet, the results will not be universal.

---

19 Torreif Dahlinn's Website on Google Scholar, https://scholar.google.com/citations?user=wW5BObEAAAAJ&hl=en, provides several especially instructive references about electrical prospecting, such as: http://lup.lub.lu.se/search/ws/files/5624610/4064852.pdf.
20 http://www.geotomosoft.com/downloads.php. Look for the document called "Lecture notes on 2D & 3D electrical imaging surveys".
21 https://www.researchgate.net/publication/269333784_Depth_of_detection_in_resistivity_surveys_A_case_study_of_a_resistive_buried_target.

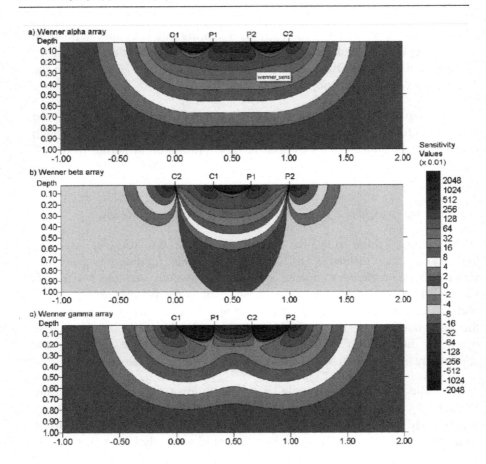

Figure 2.13. *An excerpt from Loke's tutorial (www.geotomosoft.com/) showing the sensitivity of Wenner arrays. For a color version of this figure, see www.iste.co.uk/florsch/geophysics1.zip*

2.7. Electrical resistivity tomography

We will first define electrical resistivity tomography (ERT), then show some sample direct calculations carried out with the (free) program RES2DMOD, and then invert the data obtained with these direct calculations with the demo version of RES2DINV.

2.7.1. *ERT arrays and sequences*

There are free programs for ERT arrays[22] (see also Chapter 6, the last chapter of this book, which is dedicated to the available softwares), but their use is less immediate.

ERT arrays are designed to explore the subsoil along a profile, following the x-axis, and the vertical, following the z-axis, in order to build a *vertical section*[23].

Let us consider, for example, a $C_1P_1P_2C_2$ Wenner array whose interelectrode distance is (a). Let us move it parallel to its direction, for example evenly by the distance $\Delta x = a$. This experiment yields a *resistivity profile*. Its depth of investigation could be defined as $C_1C_2/5$, by default (for example). However, we know this notion is tricky. On the other hand, we can legitimately *represent* the profile obtained below the surface, at a conventional distance from this surface, for example at the depth $z = a$. This is only one way of representing the data. In any case, we do not obtain the actual resistivity at this depth.

Afterwards, let us repeat the same process, this time with double the distance between the electrodes (2a) and, why not, with the same interval $\Delta x = a$ (or with a double interval $\Delta x = 2a$, which may be justified). As the distance has been doubled, our array basically sees twice as deep (once again, this depends on the soil itself).

With these two profiles, related to two distinct depths, we have started to create a "resistivity pseudo-section". It is a "pseudo" section because we only obtain apparent resistivity values and, especially, because the profiles are not linked to an actual depth.

The measuring process can go on, with an interelectrode distance of (3a), etc. At one point, the question of geometric coefficients, instrument sensitivity, etc., leads us to increase the interval Δx to (2a), (3a), etc.

---

22 See, for example, https://www.researchgate.net/publication/235978635_Boundless_ Electrical_Resistivity_Tomography_BERT_v_20_Open_Access_Software_for_Advanced_a nd_Flexible_Imaging and http://www.resistivity.net/?bert, as well as http://eidors3d. sourceforge.net/.
23 Geographers, geomorphologists and archeologists often prefer using the term "transect"; see, for example, http://www.hypergeo.eu/spip.php?article60.

However, we will always keep an interval that is smaller than the interelectrode distance, otherwise we will risk undersampling the variations in resistivity.

The whole process creates a pseudo-section, as is shown in Figure 2.14. We should point out that this is a pseudo-section in 2.5D, i.e. with a 2D structure but with point current sources – therefore in 3D – which suitably represent actual electrodes.

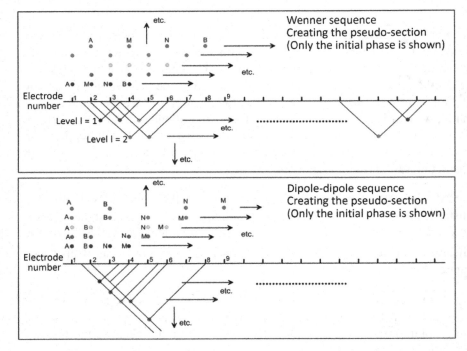

Figure 2.14. *The approach using ERT arrays. Above, the Wenner array. Below, the dipole–dipole array. The goal is to fill the pseudo-section by combining a translation (exploration along the horizontal axis) with the gradual increase in size of the array (exploration along the vertical axis). For a color version of this figure, see www.iste.co.uk/florsch/geophysics1.zip*

A fundamental element in the creation of a tomographic image is the *sequence*, which describes the position of the electrodes involved measurement by measurement. It helps the graphical creation of the pseudo-section, which is the first representation of the data measured.

Here is a sample sequence for a Wenner array:

Sequence for a Wenner array, with 20 electrodes and an initial interelectrode distance (a) used to calculate the geometric coefficient (which, however, is not shown in this table, which illustrates instead "which electrode does what"). The $C_1P_1P_2C_2$ columns are related to electrode numbers, but they may also be related to abscissas along an (x) axis (in which case, the starting point would be 0 and the following point (a).

Level	No. of measurement	C_1	P_1	P_2	C_2	V	I	K	RHO_app
1	1	1	2	3	4				
	2	2	3	4	5				
	3	3	4	5	6				
	4	4	5	6	7				
	...	...	...	...	...				
	...	...	...	...	...				
	17	17	18	19	20				
2		C_1	P_1	P_2	C_2				
	1	1	3	5	7				
	2	2	4	6	8				
	3	3	5	7	9				
	4	4	6	8	10				
	...	...	...	...	...				
	...	...	...	...	...				
	14	14	16	18	20				
3		C_1	P_1	P_2	C_2				
	1	1	4	7	10				
	2	2	5	8	11				
	3	3	6	9	12				
	4	4	7	10	13				
	...	...	...	...	...				
	...	...	...	...	...				
	14	11	14	17	20				

And so on.

In this first example, the quadripoles $C_1P_1P_2C_2$ are moved by a unit (a), regardless of the C_1P_1 interval. We can use a variant such as the sequence:

Variant for a Wenner array: movement in relation to the level. It offers fewer repetitions, but it is quicker to implement.

Level	No. of measurement	C_1	P_1	P_2	C_2
1	1	1	2	3	4
	2	2	3	4	5
	3	3	4	5	6
	4	4	5	6	7
	...	...	...	...	...
	...	...	...	...	...
	17	17	18	19	20
2		C_1	P_1	P_2	C_2
	1	1	3	5	7
	2	3	5	7	9
	3	5	7	9	11
	4	7	9	11	13
	...	...	...	...	...
	...	...	...	...	...
	14	13	15	17	19
3		C_1	P_1	P_2	C_2
	1	1	4	7	10
	2	4	7	10	13
	3	7	10	13	16
	4	10	13	16	19

Finally, here is a sample sequence for a dipole–dipole array. The latter type of array is commonly used since it offers good sensitivity, especially to surface structures, and is easy to employ.

Sequence for a dipole–dipole array. Its specific feature is that it can leave C_1C_2 unchanged, whereas we move P_1P_2 for the measurements.

Level	No. of measurement	C_1	C_2	P_2	P_1	n
1	1	1	2	3	4	1
1	2	1	2	4	5	2
1	3	1	2	5	6	3
1	4	1	2	6	7	4
	...	...	...	...	...	
1	16	1	2	18	19	16

Stop! Let us recall that the geometric coefficient of the dipole–dipole array is $\pi an(n+1)(n+2)$. This is expressed as n^3. For $n = 5$, it is equal to around 660a. Let us compare this to the Wenner array: 6.28a. This means that the signal is more than 100 times weaker than the one obtained with the Wenner array. Therefore, for $n = 16$, with a dipole–dipole array, the signal fades quickly as the distance between C_2 and P_2 increases (the "na"). Thus, we only consider small values of n, and we rarely go beyond 5 or 6. This is not a serious problem, as we will increase the offset between the electrodes, moving from (a) to (2a), then (3a), etc. Let us show this with n limited to 4, which is actually enough.

Level	No. of measurement	C_1	C_2	P_2	P_1	N
1 (distance a)	1	1	2	3	4	1
2	2	1	2	4	5	2
3	3	1	2	5	6	3
4	4	1	2	6	7	4
1	5	2	3	4	5	1
2	6	2	3	6	7	2
3	7	2	3	8	9	3
4	8	2	3	10	11	4
1	9	3	4	5	6	1
...	...	..	..	..	..	..
Last for distance = a	68	17	18	19	20	1
2 (distance 2a)	69	1	3	5	7	1
4	70	1	3	7	9	2
6	71	1	3	9	11	3
8	72	1	3	11	13	4
2	73	3	5	7	9	1

And so on.

2.7.2. *Using the programs RES2DMOD and RES2DINV*

On his Website http://www.geotomosoft.com/downloads.php, Loke provides a free program used to carry out direct calculations, RES2DMOD, which can predict a pseudo-section based on a given model. Other programs available online can carry out these calculations in 2D or 3D.

Let us provide an example of how we can use RES2DMOD, which can be opened by clicking on the link above.

Once the program has been installed, let us open the program and then load an existing model, for example one with two blocks:

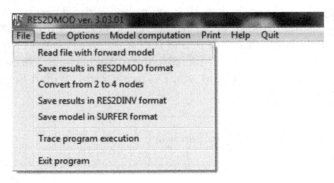

The characteristics of the model are displayed (we can easily build our own models with a simple text editor).

```
RES2DMOD ver. 3.03.01 - C:\Program Files (x86)\Geotomosoft Solutions\Res2dmod\block_two2.mod
File   Edit   Options   Model computation   Print   Help   Quit
Model name   -   Two Blocks 2 m. apart.
Number of electrodes is 37.
Number of apparent resistivity levels is 10.
No water layer
Electrode spacing is     1.0.
No remote electrodes.
Model grid depths specified in model file.
Number of model resistivity values is 3.
The number of nodes per electrode spacing is 4.
   10.00 100.00 500.00
Number of grid rows in model is 14.
Depth values for model row grid lines are :-
    0.15    0.30    0.50    0.75    1.00    1.55    2.15    2.82    3.55    4.36    5.24
    8.00   15.00   25.00
Depth grid lines extended to 15 lines.
Depth grid lines extended to 15 lines.
Wenner Alpha array
Reading of data file is complete.
```

We can "view" the model by clicking on Edit/Display model:

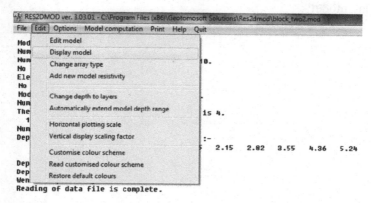

This function displays the model. However, we can skip this step and carry out the direct calculations straight away. We can do this by clicking on "Calculate potential values":

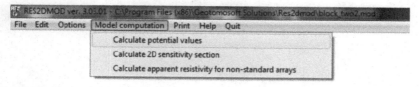

This shows the model with its grid (Figure 2.15).

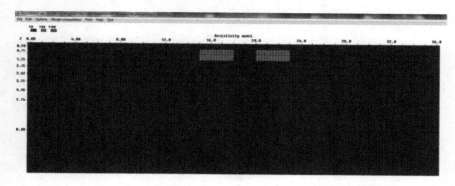

Figure 2.15. *Model obtained in a test with RES2DMOD*

Finally, we simulate the pseudo-section, which is consistent with what we would have obtained in the field, in the presence of such structures, and in this case with a Wenner array. We can do this by clicking on "Display model":

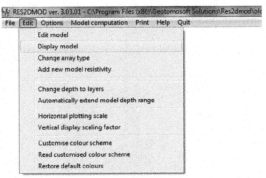

This function allows us to choose the iso-contours of the representation. The log, offered by default, works very well, as it distributes the colors according to a scaling law.

Finally, we obtain the result, as shown in Figure 2.16:

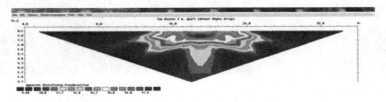

Figure 2.16. *The result of the modeling process related to the model of Figure 2.15. With this direct problem, we simulate what we would have observed in the field if this two-block structure had been present. For a color version of this figure, see www.iste.co.uk/florsch/geophysics1.zip*

This model, chosen among the examples, is made up of two blocks, whose resistivity is 100 Ωm, in an enclosure with a resistivity of 10 Ωm (in reality, these are infinite structures perpendicular to the sheet of paper, as in this case we are in 2.5D). The array is a Wenner alpha array. The image of the pseudo-section, as mentioned above, shows an apparently unique resistant structure with a complex shape. We cannot recognize our structures in this pseudo-section. What is worse is that when there are two distinct blocks, we can only seem to detect a poorly understandable shape. Let us point out that the z axis is noted as "Ps.z", which stands for "pseudo-section": these are not actual depths, just the type of representation.

Let us keep these results by saving them as a .dat file with the corresponding function in RES2DMOD.

This first example shows how hard it is to interpret the results of the measurements *directly*. The pseudo-section does not accurately represent reality and the structures are somewhat blended, due to the fact that the currents, which "carry" the electrical potential of the sources, follow a complex and vague path. We firmly need to invert the data, namely *calculate* a structure that yields the same pseudo-section as the measured one.

The program RES2DINV is an example of a program designed for inversion, and it is undoubtedly the most used at the moment. However, it requires a license that costs more than 3000 USD. There are a few free programs used for tomography (see Chapter 7), but we prefer to provide examples related to RES2DMOD and RES3DMOD due to how easy they are to use. Their demo version can introduce us to inversion and even allow us to obtain a few results. The possibility of saving files is limited (but we can always take screenshots), the iteration number is limited to three (in general, we need to go up to four or five), and, among other things, we cannot take topography into consideration (when it is not planar).

RES2DINV can do several things, especially in terms of internal parameterization of the calculations. As is often the case in geophysics, the results of the inversions depend on the choices made for these parameters. It would take us too long to explain them in detail, so we refer to the aforementioned "Course notes" provided by Loke.

To invert tomographic data, we should save it in the recommended format, which is explained in the help section of the program (also available in the demo version). The program RES2DMOD can export files directly in this format, which must be saved as .dat.

Let us consider the example of the data provided in the annex RES2DMOD. During the calculations, we have set the measurement noise at 1%, corresponding to the smallest error rate that could be possibly found in practice.

Let us move on to the inversion. We need to download the free demo version of RES2DINV from http://www.geotomosoft.com/downloads.php, install it, and then open it. In the "File" menu, we will open a ".dat" file previously created with RES2DMOD. Afterwards, let us click on "Inversion/Carry out inversion".

The result of the inversion is displayed, as shown in Figure 2.17.

Let us consider another example, i.e. the "thin dyke" shown in Figure 2.18. The "dyke" part is considered at $50\ \Omega\,\mathrm{m}$. Its enclosure is at $1,000\ \Omega\,\mathrm{m}$, while there is a cover at $300\ \Omega\,\mathrm{m}$. Geologically, this may represent an altered volcanic intrusive dyke or a water-saturated fault area in a granite enclosure under the weather layer of the granite or any other kind of relatively conductive material.

Once again, the link between the pseudo-section and reality is not evident. We can certainly see the relatively conductive "cover", but the central area seems much wider and does not seem to extend vertically. The enlargement may be associated with the idea that the currents "look for the conductor", as we often say, but in reality, this width also depends on the size of the array used, here a Wenner beta array (or a dipole–dipole with $n = 1$). The array is around 15 m long and is affected by the conductor at a distance that is at first in the range of the size (15 m) plus the depth of the source (10 m), giving to the image of the conductor an apparent width of around $15 + 10 = 25$ m.

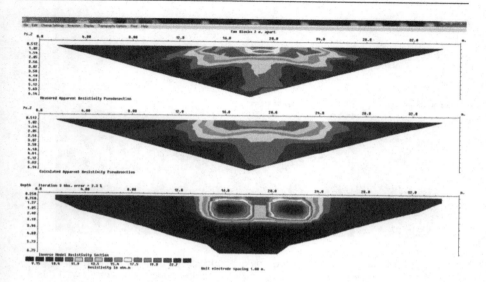

Figure 2.17. *An inversion carried out with RES2DINV. This figure is divided into three parts. Above, we can see the measured pseudo-section. Below, we can see the inverted section, and therefore the geological model (the geoelectrical model, to be precise) that would yield the same result as the actual measurements in terms of data. In the middle, we can see the pseudo-section recalculated based on the model obtained. In a way, this is the summary data corresponding to the model shown in the picture below. The quality of the reconstituted pseudo-section is limited by the number of iterations allowed by the demo version, i.e. only 3. Despite this constraint, we can see quite accurately the notion of the two initial blocks in the inverted image! The error shown (for three iterations) is equal to 2.34%. This is the average relative error between the data pseudo-section and the pseudo-section we have been able to rebuild through calculations based on the model obtained precisely following the criteria of maximum error reduction. In practice, we rarely obtain anything better than 2%, especially because reality is always more complex than the models. For a color version of this figure, see www.iste.co.uk/florsch/ geophysics1.zip*

Let us invert this model with a 5% of noise. This may seem a significant amount of noise, but we should keep in mind that we are not always dealing with the measuring noise related to the instruments (due to the precision of the voltmeter and the ammeter, etc.). This noise may have to do with what geophysicists classify as "geological noise". It concerns the effects of the heterogeneities of the subsoil that cannot be resolved by the array used (heterogeneities smaller than the interelectrode distance and 3D effects, whereas the inversion is carried out with the hypothesis of 2.5D). Geological noise includes a bit of everything. The result is provided in Figure 2.19.

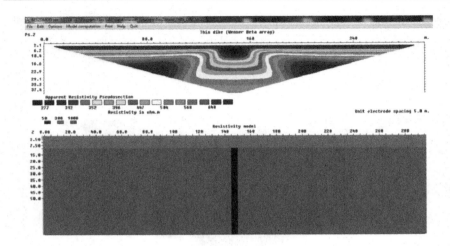

Figure 2.18. *Below, a model of porous dyke (conductive) under a cover of average resistivity in a resistant enclosure; above, the pseudo-section associated with this model. For a color version of this figure, see www.iste.co.uk/florsch/geophysics1.zip*

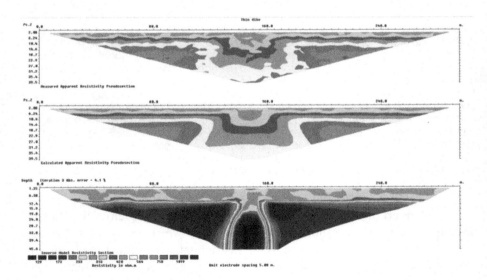

Figure 2.19. *Inversion of the noisy pseudo-section (we can clearly see the noise!). We can observe in the inverted image that the superficial layer and the enclosure are correctly recovered, with resistivity values close to those transmitted to the model. This does not hold true for the vertical structure. If it is clearly present and established like in the model, it widens toward the bottom and its inverted resistivity is stronger than in the original. For a color version of this figure, see www.iste.co.uk/florsch/geophysics1.zip*

This case illustrates particularly well the loss of resolution of the electrical method as the depth increases.

This first experiment carried out with RES2DMOD and RES2DINV shows simultaneously the strength of "direct-current" geophysical methods and their limitations. The reader can analyze several other scenarios thanks to the numerous examples related to RES2DMOD. No two cases are alike, and it is the experience gained with the program and electrical methods, together with previous knowledge about geology, that allows us to ensure the (always fairly relative) reliability of the result of such inversions.

There are countless examples of tomographic images online, which can be found with the keywords "Electrical Resistivity Tomography".

Actual tomographic examples are provided in Chapter 4, which is dedicated to induced polarization (IP).

At this stage, this program (as well as other programs) provides good images to the extent that we need to mention how these results are obtained, i.e. to provide some theory, which readers are free to skip (it is in no way necessary to master these theories in order to continue reading this volume).

The theory of direct current electrical prospecting

This theory first relies, on the one hand, on Maxwell's equations(#), and, on the other hand, on Ohm's constitutive law $\vec{J} = [\sigma]\vec{E}$ ($\vec{J}$ is the current density in A/m², $[\sigma]$ the conductivity tensor, whose dimensions are 3X3, which becomes a simple real number (a "scalar") in isotropic environments, in S/m, and $\vec{E}$ is the electrical field, in V/m). As we are dealing with direct currents, the functions do not depend on time, so we can eliminate the temporal derivates from Maxwell's equations. After a few changes, assuming that there is a point current source I in the shape of a Dirac delta function, we obtain the equation by writing that "nothing is lost, nothing is created" (conservation law), this being applied to the charged particles (ions and/or electrons). Besides, all the current transmitted to an electrode is recovered by the other. (only some energy at work is transformed into heat (Joule effect)).

Mathematically, for the potential V, we can write this equation in a Cartesian system (0 x, y, z):

$$\vec{\nabla} \cdot \left([\sigma]\overrightarrow{\nabla V} \right) = -I\delta(x, y, z).$$

$\bar{\nabla}$ represents the "nabla" operator(#).

The problem is to solve this equation in V (to find V) for a spatial distribution with a given conductivity. Let us point out that two objects are idealized and not "physical": the Diract delta function(#) and the potential V itself as it is defined more or less like a constant, and only potential differences can be found empirically. Moreover, what is measured and what allows us to deduce a potential difference is *always* a charge movement. The potential is only a practical mathematical notion used to carry out calculations.

We should add some "boundary conditions" to this equation, otherwise we risk not being able to solve it. These conditions are simple: we assume that the potential approaches zero toward infinity (we could choose any value for the potential to infinity) and that no current can cross the surface of the earth (and spread in the air).

Making use of the problem posed by the equation and these boundary conditions, an expert in numerical analysis can solve these equations by using tools like finite differences or finite elements(#). Readers interested in these questions can refer to Loke's course notes.

2.8. 3D tomography

The basic notion of 3D simply involves associating 2D sections that are organized spatially. An actual example is provided in Figure 2.20 in relation to tomographic images taken at the Sutherland observatory in South Africa.

We can see here a relatively conductive layer, a regolith, that logically lies on unaltered volcanic rock. Without these few pieces of information about the geological features, we could only picture a specific stratification.

Determining the interface between the two is not immediately obvious: this is another drawback of the bad resolution of electrical methods. A transition taking place at the inflection point between resistivity and depth may correspond to a boundary of $1,000\,\Omega\,m$. Every geophysical interpretation includes a subjective component.

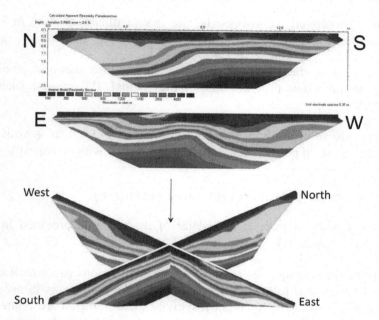

Figure 2.20. *Associating sections of the same soil represents the first step in a 3D approach. For a color version of this figure, see www.iste.co.uk/florsch/geophysics1.zip*

This first example is perfunctory: only two sections. A more efficient 3D model assumes that the 3D space is explored more in detail, so that we can finally obtain a more substantial description of the resistivity values in relation to the three axes (x, y, z).

Another method involves creating several sections close to one another. As for the data representation, there is nothing to change, as in any case pseudo-sections are only one way of representing data and not the representation of actual resistivity "as it is" (even if its quality should deteriorate because of the low resolution).

We can use two approaches. The first involves inverting the sections separately, like 2.5D models, and then representing them side by side, by looking for a perspective effect. This first method is "somewhat wrong". An anomaly seen by a 2.5D array (an ERT array) is always represented, even after the inversion, below the corresponding profile. The body seen this way may not be under the profile, but shifted laterally with respect to the profile. Moreover, we always run the risk of misinterpreting in 2.5D what in nature, with rare exceptions, is always 3D.

Besides, the correct way of inverting these close sections involves applying an inversion program that creates actual 3D images, such as RES3DINV.

We provide an example of this kind of analysis in Chapter 4, which is dedicated to induced polarization (which is associated with resistivity).

2.9. Direct current electrical resistivity mapping

Electrical mapping was at the center of the example provided in the introduction (Figure 1.3).

The principle is simple: an array with fixed dimensions and directions is moved on the ground following the perpendicular x and y axes in order to explore a given soil. The apparent resistivity measurements are related to the environment of the array to create a map. Figure 2.6 illustrates this principle by taking into consideration a team moving along x and y.

The (indicative) depth of investigation and the representative volume associated with the array are thus invariable in a mapping operation. For example, for a Wenner array, we consider an area of influence between the electrodes C_1C_2 and a depth of investigation in the range of $C_1C_2/5$. The most commonly used arrays are the pole–pole array (whose drawback is that it requires two long cables for the electrodes C_2 and P_2, but whose images are robust), and the Wenner alpha and beta arrays (the beta array is better in terms of accuracy of the anomalies in relation to the structures). Professionals also employ squared or trapezoidal arrays by towing systems behind vehicles when carrying out wide-scope agricultural or archeological surveys. The electrodes are thus points attached to wheels. The towed system generates parasites that a selective statistical process attempts to eliminate. Here, the main goal is to cover the huge surfaces.

In the agricultural domain, resistivity reveals with great accuracy the clay content of the soil, contributing to a sustainable (so-called "precision") kind of agriculture, whereas in archeology, large sites may be mapped. In these two fields, the acquisition speed goes up to 10 hectares per day with metric resolution.

Compared with electrical tomography, arrays involve the sampling of a single pseudo-depth on several surface and parallel profiles. It goes without saying that a map created with several sizes of the same array is also a vertical exploration and ultimately allows us to combine data that may be used in a basic 3D protocol[24].

An example of pure geological prospecting is provided in Figure 2.21.

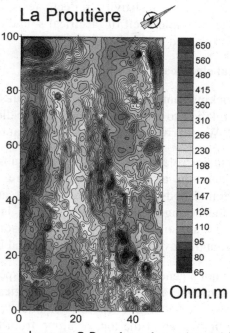

Pole-pole map C_1P_1 = 1 m, 1m x 1m grid

Figure 2.21. *An example of pole–pole prospecting in gneissic environment in Vendée (France). The ground is sloping and abuts on a stream in the north-west. The resistant areas (orange to brown) reveal the presence of unaltered granite, whereas the conductive areas (green to violet) denote a marked alteration and a permeable sandy surface when its clay content is not too high. The directions can be explained with the preferential way in which water flows, surrounding the parts that have remained untouched. The spot in the upper left-hand corner corresponds to a surface accumulation of clay, which is impermeable and corresponds to a muddy area. For a color version of this figure, see www.iste.co.uk/florsch/ geophysics1.zip*

24 https://ia600306.us.archive.org/24/items/HandbookOfAgriculturalGeophysics/Handbook%20of%20Agricultural%20Geophysics.pdf.

2.10. Vertical electrical sounding (VES, or in short, electrical sounding)

2.10.1. VES and horizontally layered ground

Figure 1.4 reminds us that the larger the array, the deeper it can "see".

In a given point, as we vary the size of an array (for example a $C_1P_1P_2C_2$ Wenner array), the apparent resistivity reflects resistivity as the depth increases. This is the principle itself of electrical sounding (here Wenner): apparent resistivity is charted in relation to a = $C_1C_2/3$.

In reality, vertical electrical sounding uses gravity.

This slightly shocking sentence recalls the significance of gravity in our lives as well as in relation to electrical sounding. If it did not exist, would we be able to walk? Would rivers flow? What about plants?

Gravity plays a major role in the structure of the subsoil. It ensures the stability of the planet on a global scale but, more locally, it leaves sediments on horizontal beds at the bottom of lakes and seas. In short, gravity originates the arrangement of the ground in layers down to the center of the Earth. As a result, in many cases the subsoil under our feet is structured as "horizontally layered ground" in horizontal beds or strata.

Naturally, this is not a general rule and it depends especially on the scale considered, topography, the erosion dynamics, geological events, etc. The way in which granites are altered[25], leaving "boulders", i.e. intact granite blocks (at the bottom or even at the top, resulting from the fine sediments transported away), near the surface, also finally involves a kind of stratification, including from the bottom up unaltered granite, fissured granite, sandy surfaces, reaching then clay sand at the surface. In a geoelectrical image focusing on depth, the resistivity will reveal the stratification more or less in terms of these kinds of heterogeneities.

Together with gravity, the atmosphere, rain and plants also contribute to the structure of the subsoil in layers that are more or less parallel to the surface.

---

25 See, for instance, https://www.researchgate.net/profile/Robert_Wyns/publication/230279 986_The_fracture_permeability_of_Hard_Rock_Aquifers_is_due_neither_to_tectonics_nor_ to_unloading_but_to_weathering_processes/links/02e7e5230608004af1000000.pdf.

Thus, except for some specific cases, the ground under our feet is often horizontally layered. The soil may vary 5 m away – we know for a fact that the subsoil is much more heterogeneous than it seems – or only 500 m farther away. If we want to find out, we need to use ERT, carry out electrical mapping or employ other geophysical methods. In many cases, our goal is to detect the differences in this horizontally layered structure, as is the case for archeological remains or when we are looking for a fault hidden by more superficial layers.

The fact remains that this horizontally layered structure can be found frequently, justifying the choice of "vertical electrical sounding", which is the quintessential 1D type of analysis, deployed toward the vertical dimension.

The Wenner, Schlumberger ($C_1P_1P_2C_2$, but with a small P_1P_2 compared with C_1C_2), dipole–dipole and pole–pole arrays are the most commonly used.

A horizontally layered type of ground may include 2, 3, 4, 5 (or more) geoelectrical layers. Consequently, electrical sounding presents variations that reveal this series. However, things are more complicated than they seem. The interpretation of electrical sounding faces all the traps of geophysics. It is even paradigmatic in this respect. However, before focusing on these traps, let us describe in more detail the notion of sounding itself.

Figure 2.22 illustrates an electrical sounding involving three layers. The demo version of the software used to calculate this result is available and mostly sufficient for our needs: IPI2WIN (three capital Is)[26]. First, we need to reinput an experimental data set by clicking on File/New VES points. There are several ways of encoding it, but it is enough for us to encode the sounding directly as a_i (interelectrode distances). Otherwise, we can encode the potentials and the currents. With a Wenner array, we have $C_1C_2 = 3a$, which can vary, and some $\rho_{a,i}$ values that represent the apparent

---

26 Another free-trial program called IX1D can be found here: http://www. interpex.com/ix1d/ix1d.htm. Like IPI2WIN, it allows us to move layers by clicking on them and quickly learn about how soundings can be interpreted. There are also other programs, which can be found online.

resistivity measurements for each a_i. The experimental curve is plotted automatically.

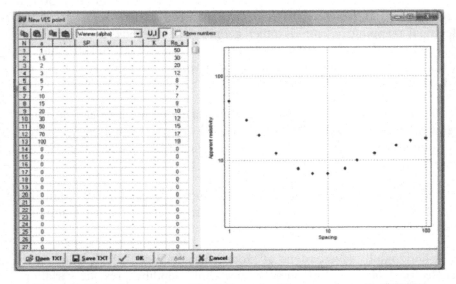

Figure 2.22. *Data coding and the first graph of the sounding with IPI2WIN*

Afterwards, we save these points. The program proposes a *model*. Figure 2.23(a) shows the experimental points (black curve) and offers here a two-layer model based on the beginning and the end of the electrical sounding. The program plots this geological model (in blue) and the resulting curve (sounding simulated for this two-layer model) in red.

The columns of the table represent the apparent resistivity, the thickness of the layers, the depth of the interfaces (the sum of the thickness of the first layers) and the altitude of the interfaces.

This is certainly not satisfactory, so we will require a small human intervention to re-enter the "three-layer-model" hypothesis that we are considering.

The "Point/New Model" function offers an alternative model (Figure 2.23(b)).

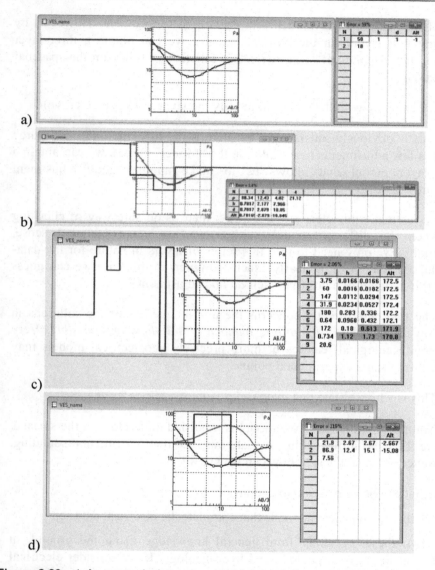

Figure 2.23. a) An excessively summarized interpretation, considering a two-layer model. b) The model automatically offered by the program. c) With a model that includes several layers, we obtain an adjustment in relation to the experimental data that is very satisfactory, but not even remotely realistic. What is worse, some layers have no influence within the range of the sounding (between $C_1/C_2 = 1$ and 100 m). d) After the two-layer model, which was unsuitable, we have another certainty: the three-layer model shown here cannot be the right one. The theoretical curve is not consistent with the data. For a color version of this figure, see www.iste.co.uk/ florsch/geophysics1.zip.

This is a four-layer model with remarkable adjustment. However, by joining two layers (in the "Model" menu), we can obtain a three-layer model that, despite not being perfect, is satisfactory (click on the spacebar to readjust).

We can play with the mouse to modify the layer model (blue curve).

This is obviously satisfactory, but here is the first trap: why not more? With a few adjustments (use "split" in the "Model" menu), we can obtain a nine-layer model that allows us to make another good adjustment (Figure 2.23(c)).

Let us make this clear straight away: there is no sure way of choosing between the different models, unless we use some common sense. Naturally, we only need to find this series of layers in nature for the data-model association to be nearly exact. However, in this case we can guess that it is a matter of a do-it-yourself *ad hoc* adjustment.

The trap here directly concerns the non-uniqueness of the solutions in the inverse problem. *An infinite number of solutions are possible.* Very different geological features or more precisely geoelectrical models may provide the same experimental points.

This is why we state that *geophysics cannot work on its own.*

Naturally, in this case, the model is modified quite close to the surface, where it has no influence in the scale range of the experimental sounding. However, in general cases, how can we know?

What can be an addition to geophysics?

– Other geophysical methods that complete the electrical data set;

– hypotheses resulting from general knowledge about the ground, in particular in terms of geology and hydrogeology. If we consider electrical methods, we may have an idea about the resistivity of the ground and/or the nature of the beds, so that we can avoid doing something stupid. If available, drilling data is very useful and often allows us to set a specific parameter;

– practice and experience, thanks to which we do not end up twisting data to suit our own projective, if false, expectations.

This refers to what we said about geophysics and inversion: they spell out what is not possible. The model of Figure 2.23(c) is possible but highly unlikely. On the other hand, the following model, in Figure 2.23(d), shows a soil *that is not compatible with the data*. Therefore, it is *certain* that we will not find under the electrodes a horizontally layered type of ground with a relative resistant element viewed laterally and as thick as the one shown in the figure.

Geophysics does not have the monopoly on this type of analysis by exclusion, as astronomers and doctors often adopt the same approach. From an epistemological point of view, this approach should be undoubtedly linked to the definition of science given by Karl Popper[27]: what is scientific can be refuted. Geophysics does not tell us what something is, but what it is not.

This often makes us look good, but in practice, the *a priori* knowledge we have about the ground, common sense, and the application of Ockham's razor[28] lead us close to the actual structure of the ground! Moreover, there is a strong Bayesian[29] spirit in geophysical interpretation. What we mean is that *before* taking measurements, we *already* had some information; however, this led to a huge domain of possible solutions. Therefore, there is a new, restricted set of possible solutions that result from and finally match the measurements, which are used to reduce, often greatly, the initial space of possible solutions. Geophysicists may imagine all the bad solutions to interpret these experimental data; clearly, they will consider fewer solutions that fit the experimental data.

There is no universal solution for the inversion or interpretative process related to electrical soundings as well as mapping and panel arrays. The best tool on which geophysicists can rely is first of all experience as well as

---

27 https://en.wikipedia.org/wiki/Karl_Popper.
28 https://en.wikipedia.org/wiki/Occam%27s_razor.
29 https://en.wikipedia.org/wiki/Bayesian_statistics.

some knowledge about the existing traps. These elements may even represent what defines the job.

Let us explore some more the traps involved in electrical soundings, but first let us consider the question of logarithms.

2.10.1.1. *Why do we work with logarithms?*

The question of logarithms is fundamental. It regards both resistivity and the size of the array. It results from multiplicative scaling laws that can be easily understood.

Let us start with the geometric scale. Let us imagine a resistant or conductive layer, a meter thick, situated in homogeneous ground. Let us assume that this bed is 1 m deep and thus occupies the ground between 1 and 2 m. We can see that a sounding will easily "see" this bed; it is enough to vary the C_1C_2 of the Wenner array between 0.1 and 20 m.

Let us assume that the same bed is situated hundreds of meters underground. It is now merely a 1 meter bed 100 m deep, and it is quite tiny in the vertical layout of the soils. We will need a large array to "find" it, but most likely it will be nearly invisible. In fact, it will be a bed 100 m thick and 100 m deep, occupying the portion from 100 to 200 m, which would correspond to our 1-m thick and 1-m deep bed. It is by increasing the electrical sounding by a factor of 100 that we can maintain the scale ratio. In a log-log chart, it is a simple two-decade translation (we refer to the "modulus" of a logarithmic scale).

This is a well-known property, illustrated by a counter-example considered in J. L. Aster's[30] work, provided in Figure 2.24 (if the ratios had been kept, the curves would have simply been translated).

Thus, for a given geometric model, the fact of multiplying all the geometry by a given factor will preserve the relative influence of each bed. On a "log" scale, if we multiply by a number greater than 1, the sounding will simply shift to the right; otherwise, it will shift to the left. This is due

---

30 J. L. Astier, 1971. *Géophysique Appliquée à l'Hydrogéologie*. This is an excellent French book, unfortunately out of print. A recent book in English is the one by R. Kirsch, *Groundwater Geophysics*, Springer, 2006.

to the fact that the log multiplication, after the transformation, is equivalent to a translation (as $\log(Ax) = \log(A) + \log(x)$).

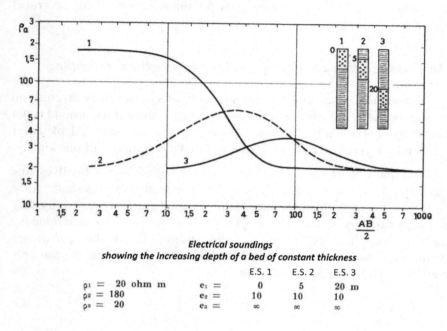

Electrical soundings
showing the increasing depth of a bed of constant thickness

				E.S. 1	E.S. 2	E.S. 3
$\rho_1 =$	20	ohm m	$e_1 =$	0	5	20 m
$\rho_2 =$	180		$e_2 =$	10	10	10
$\rho_3 =$	20		$e_3 =$	∞	∞	∞

Figure 2.24. *The increasing depth of a 10m-thick bed involves the decrease in its relative significance. The multiplicative scaling law unsurprisingly does not apply (drawn from Astier, 1971)*

Let us consider the question of resistivity. Let us analyze a given ground, for example a series of three layers with resistivity values equal to (ρ_1, ρ_2, ρ_3). Let us multiply all the resistivity values by the same factor, let us say B, to obtain $(\rho_1, \rho_2, \rho_3) = (\rho'_1, \rho'_2, \rho'_3) = (B\rho_1, B\rho_2, B\rho_3)$. The resistivity ratios between the soils will be the same. Moving on to the log, the sounding will be translated upwards if $B > 1$ and downwards if $0 < B < 1$. By translation, the curves of the soundings can be superposed.

Thus, we obtain here a *scaling law* that is frequently used in geosciences, where an electrical sounding will not change its shape (which can be superposed by translation) due to scaling and resistivity homothety(#). If we consider the previous sounding and we multiply all the thickness values by a given factor, we obtain then a superimposable curve (horizontal translation). The same goes for the resistivity values (vertical translation).

2.10.2. *Difficulties and traps involved in electrical sounding*

– The sounding may look like the sounding of a horizontally layered soil and may be easily adjusted as if this were the case, even if the ground is not necessarily horizontally layered. Taking a measurement adapted for horizontally layered ground does not mean that the ground is of this kind.

– The question of the *equivalences* is very significant: it involves fine beds (this is a relative aspect), whether conductive or resistant. As a conductive fine bed channels the current, an increase in its resistivity may be compensated by an increased thickness. Thus, we can only determine the ratio resistivity/thickness (or thickness/resistivity). This is the *equivalence law* for a conductive bed, which can be defined by saying that we can only determine its *longitudinal conductance*:

$$C_L = \frac{e}{\rho}.$$

In other words, a bed of variable thickness $\{e, e', e'', e'''...\}$ and similarly variable resistivity $\{\rho, \rho', \rho'', \rho'''...\}$ that, while remaining thin (as a reference, let us say at least four times less thick than the underlying and overlying beds) verifies:

$$\frac{e'}{\rho'} = \frac{e''}{\rho''} = \frac{e'''}{\rho'''} = ... = C_L,$$

will yield indiscernible experimental electrical soundings.

Similarly, the equivalence law of a resistive bed, which will have to be transversally crossed by the current, will be written as:

$$e\rho = R_T.$$

This is the transversal resistance, because of which an increase in thickness (e) can be compensated to a certain extent (until our bed is no longer thin, roughly when the bed will be thicker than ¼ of the thickness of the enclosing beds).

These questions of equivalence become complex when there are several alternating resistant or conductive beds. At most, a series of alternating very conductive and very resistant thin beds will yield a flat sounding. However, in such cases we have to resort to anisotropy (introduced further on).

– The *disappearance phenomenon* takes place in beds of average resistivity. For a series of soils such as $\rho_1 > \rho_2 > \rho_3$ (and, inversely, with $\rho_1 < \rho_2 < \rho_3$) the second layer may be removed from the model without changing the sounding. This happens again in a multilayer sounding.

– The question of *sensitivity* arises for all arrays, both in relation to soundings and mapping, but maybe it is with ERT arrays that the question becomes more...sensitive. This question may be formulated as follows: if we hypothesize that a given structure (a bed, a buried structure, anything) is present, does this structure *actually* affect our measurement? Geophysicists ask: "Can we see this structure?". Some inversion programs edit this sensitivity as a section. Thus, we can find out if there are blind areas, which represent in most cases places where current hardly penetrates; the worst example involves resistant cavities, therefore a conductor that would be invisible on the inside (sensitivity inside a cavity is therefore equal to zero, but since programs do not take into consideration cavities unless we specify so, they may provide a sensitivity not equal to zero in this area, which would depend only on the geometry of the array).

– Let us briefly tackle the question of *anisotropy*(#). Anisotropy is frequently found in nature, especially in significantly stratified sediments, where we distinguish between transversal and longitudinal properties. We refer to transverse resistance ρ_T and longitudinal resistivity ρ_L.

Longitudinal resistivity is always weaker than transversal resistivity, as we can clearly see if we imagine a stratification made up of alternating very conductive and very resistant beds. Longitudinally, the beds are parallel and play the role of parallel conductors. Transversally – namely, perpendicularly – the resistant beds will prevent the current from flowing.

The *anisotropy coefficient* and the average *resistivity* are by definition:

$$\lambda = \sqrt{\frac{\rho_T}{\rho_L}} \text{ and } \overline{\rho} = \sqrt{\rho_L \rho_T} \text{ , respectively.}$$

The anisotropy coefficient in natural environments ranges from 1.3 to 2, going from stratified sands to fine alternating banks of marl with variable clay content.

In an anisotropic homogeneous environment, the equation that yields the potential in isotropic environment in the half-space, produced by a surface electrode (namely, $V = \dfrac{\rho I}{2\pi r} = \dfrac{\rho I}{2\pi\sqrt{x^2 + y^2 + z^2}}$), must be replaced by:

$$V = \frac{\rho_L \lambda I}{2\pi\sqrt{x^2 + y^2 + \lambda^2 z^2}} \, .$$

At the surface, therefore for z = 0, the equation is reduced to $V = \dfrac{\overline{\rho} I}{2\pi r}$.

We cannot distinguish this value from the case of an *isotropic* homogeneous ground, whose resistivity would be $\rho = \overline{\rho}$. As a result, it is impossible to find out whether a homogeneous ground is isotropic or anisotropic.

We can see that not taking into consideration anisotropy in an electrical sounding results in thickness values found by excess. An independent piece of information, for example related to drilling, allows us to find out the excess factor and determine the anisotropy factor.

Naturally, researchers have made progress and adapted several methods to overcome this difficulty, but here we cannot describe all these tricks. One of the bibliographies provided by Loke[31] includes several references, some of which focus on how anisotropy should be regarded in certain cases.

Finally, the theory of electrical soundings includes many notions used by traditional geophysicists. "This would require a book". The fact is that

---

31   https://www.researchgate.net/publication/260723789_Recent_developments_in_the_direct-current_geoelectrical_imaging_method.

there are already quite a few: readers should use a search engine dedicated to books with the keywords "geoelectric sounding" for evidence.

2.10.2.1. *How to avoid the traps?*

First of all, we will avoid these traps by being fully aware of this type of ambiguity and difficulty, aware of equivalences and disappearances, and also aware of the fact that the electrical method, like many others, works mainly by eliminating some possibilities rather than by revealing some concrete facts.

A general approach, which is not always easy, especially for a decision-maker who must make a choice based on geophysical results, involves working in terms of probabilities. This soon becomes a quagmire, as we would have to handle too many hypotheses and possibilities. Once again, we can always rely on common sense, experience, Ockham's razor and especially a certain dose of humility that will prevent us from turning our wishes into realities (especially if our wishes match the data).

There are also some "tricks" used as a sort of "quality control".

– During an electrical sounding, carrying out two perpendicular soundings allows us to exclude the possibility of a horizontally layered soil if the soundings differ. Unfortunately, the opposite is not true: two absolutely identical cross soundings do not imply anything, other than the fact that the underlying structures lead to this situation. Once again, we have the possibility of eliminating rather than confirming hypotheses.

– In 2D tomography, an anomaly could come from the side, namely be determined by a structure that is not directly under the profile but shifted. There is only one solution: carrying out parallel soundings at distances in line with the depth of investigation of the panel array used.

How do we calculate an electrical sounding in horizontally layered ground? The basic equation is always the one mentioned in section 2.7.2, in the box entitled "The theory of direct current electrical prospecting", namely

$\vec{\nabla} \cdot \left([\sigma] \overline{\nabla V} \right) = -I\delta(x, y, z)$. There is a significant simplification: in horizontally

layered ground, there is cylindrical symmetry with a vertical axis through the electrode. From 3D, this mathematical problem becomes 2D in potential according to the radius and the depth (r.z), and resistivity – therefore, the curve of the sounding too – depends only on one parameter, i.e. the depth z. We can see that the potential at the distance x from the source, close to the surface, uses a "Hankel transform"(#) of this type:

$$V(x) = \frac{\rho_1 I}{2\pi x}\left[1 + 2x\int_0^\infty K(\lambda)J_0(\lambda x)dx\right],$$

where J_0 and K represent the Bessel function of order 0 and K is a "node" that depends on the stratified environment. There are different ways of calculating K and this integral. Readers can find the complete theory in a book by Koefoed[32], which can be borrowed from the library.

2.10.3. *Carrying out vertical electrical sounding*

VES is adapted for horizontally layered ground (horizontally stratified). The first question to consider is to see if we can reasonably make this hypothesis. Two cross-soundings, even if they cannot confirm that the ground is horizontally layered, will allow us, as we have already pointed out, not to discard the sounding. Naturally, there will always be some differences between the two soundings, but everything is a matter of degree.

When we explore a certain surface, the question of horizontally layered ground may arise on small and large scales or, in other words, from the superficial part of the sounding to its deep part. For the same geological configuration, we can obtain a sounding whose top is horizontally layered (a humus layer and then uniform sandy soil) and whose bottom is not (deep lateral variations in the alteration of granite). The opposite may also be true; there may be surface heterogeneities but a regular alteration horizon in depth. In the former case, the beginning of the cross-soundings (which may also be carried out in different points) will be similar, whereas in the latter, the ends of the curves are similar.

It is useful to do some electrical mapping (to use an array) before carrying out a VES, in order to find out the degree of heterogeneity of the ground. This allows us to carry out the sounding in the most useful areas, i.e. those that minimize lateral variations.

2.10.3.1. *Choosing the array*

During an electrical sounding, a measurement is taken from the symmetrical center of the sounding on the point called "O". This makes sense when soundings are located on a map but also for symmetrical

---

32 O. Koefoed, 1980. *Geosounding principles* (Vol. 1). Elsevier Science & Technology.

reasons. Naturally, apparent resistivity integrates the subsoil "on average", namely adding or subtracting resistivity in a volume whose dimensions are related to size of the array, and not only on this point.

– The pole–pole array is only suitable for small soundings, let us say $C_1P_1 \leq 5$ m, as we need to stretch the cables of the "electrodes to infinity" more than 20 times the distance C_1P_1. The obstacles on the ground, as well as the fact that a large P_1P_2 will involve disturbances, limit this ambition. However, the advantage of a pole–pole array is its relative lack of sensitivity to the small heterogeneities of the surface. It yields a better "average" than the other arrays.

– The Wenner alpha and beta arrays are more or less universal and very similar to each other. For large soundings, we gain a factor of 3 with the alpha array (this is due to the geometric coefficient: $K = 2\pi a$ for the Wenner alpha array vs. $K = 6\pi a$ for the beta). The Wenner beta array reveals fewer artifacts than the alpha array and should be preferred.

– The main advantage of a dipole–dipole array, with a fixed n, is that it requires fewer cables, as the lines C_1C_2 and P_1P_2 are separated and even far from each other, possibly with the transmitter on one side and the receiver on the other. On the other hand, the signal fades quickly as n increases, since the geometric coefficient is $K = \pi a\, n\,(n+1)(n+2)$ (which consequently increases like n^3).

– The Schlumberger array, equipped with a coupling system, is practically the most used. The fact of having a fixed P_1P_2, whereas only C_1C_2 increases (at least for each sounding phase) is a practical advantage. However, the "couplings" present "jumps", which are discontinuities that appear when we change P_1P_2. But these defects are useful, as they provide additional information about the presence or absence of heterogeneities. Let us describe this approach in more detail.

2.10.3.2. *Schlumberger soundings*

Schlumberger soundings, based on electrodes aligned as $C_1P_1P_2C_2$, assumes that $P_1P_2 \ll C_1C_2$. We prefer using $L = OC_1$ and $l = OP_1$ to record the data, as the center of the array can be used as a reference point $x = 0$ in the field. The geometric coefficient is equal to:

$$K = \pi \frac{(L^2 + l^2)}{2l}.$$

We can often approximate it as $K = \pi \dfrac{L^2}{2l}$, since with this array we assume "by principle" that $1 \ll L$. However, this is dangerous and honestly useless, as it does not significantly help us with the calculations, whereas in practice this condition is not completely respected. It turns out that, as C_1C_2 increases, there is a moment when the signal at P_1P_2 is too weak. Therefore, we increase P_1P_2. The nature of the ground may imply that the apparent resistivity value changes slightly for the increased P_1P_2 (whereas ideally, for a fixed C_1C_2, the apparent resistivity depends only on C_1C_2). Thus, to make this change more reliable, we set up a *coupling* system, i.e. an operation involving the readjustment of the pieces obtained for each value of P_1P_2.

Figure 2.25 (as well as Figure 2.26) below shows the sheet of a typical sounding that describes step by step what we apply in practice. After the measurements, the result is not a single sounding curve, but a set of them whose parts will be joined by hand. The goal is to obtain a continuous curve, to increase the validity of the hypothesis of a horizontally layered soil.

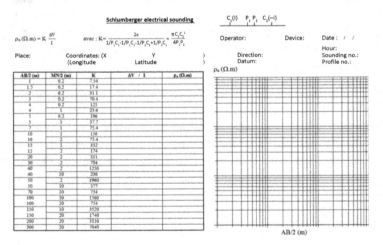

Figure 2.25. *A sample soil sheet of a Schlumberger sounding. The log-log paper allows us to plot the curve as we go along and to appreciate in real time the quality of the data*

To carry out this operation, we choose one of the parts of the curve and we join the other pieces by pure vertical translation of the whole piece of curve. Thus, the data are in a way falsified on purpose. This is for a good cause and, on a geological level, this means eliminating small changes in the soundings that derive from slight lateral variations "that could not be detected".

Sometimes, even if this is not a habit, the couplings system may be nearly perfect. If it is very imperfect, we know that there are lateral variations along the sounding. This is a piece of information: for a strictly horizontally layered soil, the couplings must be good[33].

This coupling protocol does not exist in Wenner, pole–pole or other types of arrays. In conclusion, long live Schlumberger soundings[34]. This type of sounding allows us to save some P_1P_2 measurements, detect the presence of heterogeneities and evaluate their significance.

Below, Figure 2.26 illustrates an example of sounding. Let us see how it is interpreted with IPI2WIN.

Entering the data represents the first step. Here is the corresponding window:

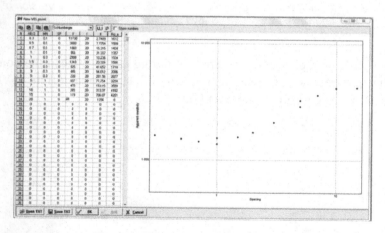

Figure 2.26. *Data for a Schlumberger sounding. The data are re-entered as V and I. Let us point out the "simple" couplings (on one point; sometimes they involve two or even three points)*

---

33 Mathematically speaking, no coupling is good, as a change in P_1P_2 represents a change in the array. However, the approximation is as good as P_1P_2 is small compared with C_1C_2.

34 Readers can read the landmark work of the Schlumberger brothers at: http://gallica.bnf.fr/ark:/12148/bpt6k64569898.

Let us point out that once we have the couplings, P_1P_2 varies as follows: $0.1 \rightarrow 0.3 \rightarrow 1$.

When the data are validated, the program plots the segments of the curves, but it also joins the pieces. As a result, we obtain a continuous curve. Here, we illustrate the method by referring to the curve's last segment, which has been taken as the reference. The result is shown in Figure 2.27.

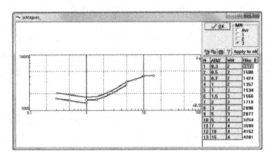

Figure 2.27. *For Schlumberger soundings, the program automatically joins the sounding segments. It is important to obtain a continuous curve if we want the effectiveness of the inversion programs to be satisfactory. For a color version of this figure, see www.iste.co.uk/florsch/geophysics1.zip*

If we click on the small green arrow pointing to the right in the control bar, the program determines a number of beds and offers a first interpretation. The result is shown in Figure 2.28.

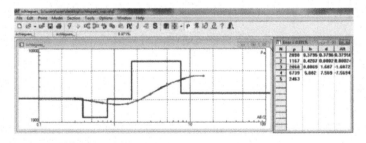

Figure 2.28. *The program offers…five soils! The programmer probably did not watch "The Name of the Rose"[35]. For a color version of this figure, see www.iste.co.uk/florsch/geophysics1.zip*

---

35 Umberto Eco's character, William of Baskerville, reminds us in particular of the aforementioned William of Ockham and his principle, which we have expressed in relation

"Join" (in the "Model" menu) allows us to reduce this slightly excessive model to a three-soil model. Afterwards, we run the inversion again (by pressing tab) or we click on the beds directly and modify them ourselves. For example, we can obtain the result illustrated in Figure 2.29.

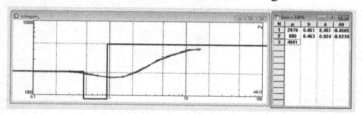

Figure 2.29. *In the lack of other information, a simpler interpretation is more likely to be more faithful to reality. The interpretation of this wooded land in a valley perched on the "granite peaks" of the Vosges Mountains, in the middle of summer, is quick: 46 cm of vegetative cover, very permeable and dried of the substances taken by trees and other ferns. Then, a layer of sandy soil of the same thickness with moisture residues; finally, the relatively unaltered bedrock, which is very resistant and situated 92 cm underground – approximately, of course. For a color version of this figure, see www.iste.co.uk/florsch/geophysics1.zip*

The error has naturally doubled (from 1 to 2%), but finally we start wondering whether this soil may not be *exactly* horizontally layered.

We can use the program to analyze the equivalence in the conductive bed, as shown in Figure 2.30.

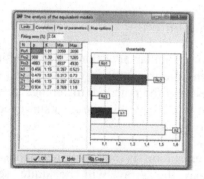

Figure 2.30. *This excellent program can perform error calculations. Rho2 and h2 are poorly determined: there is an equivalence*

to our sounding in the following terms: "we should not multiply the explanations and the causes unless strictly necessary".

A first table provides the formal errors about our beds, and with the "Correlation" tab we can find out the correlation coefficient. Finally, the "minimum" and "maximum" functions in the "Model" tab allow us to explore the possible extremes. For example, this is what we obtain by clicking on "minimum" (Figure 2.31).

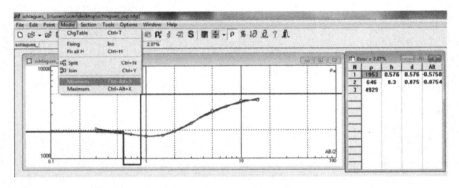

Figure 2.31. *The minimum and maximum function provides the range of possible solutions for this sounding*

With the same software, several soundings with the centers aligned on a profile allow us to create a pseudo-section (apparent resistivities) and then a section (of inverse resistivities). The number of soundings is limited to 10 in the demo version, which is often sufficient.

2.11. The rectangle method

Let us consider a rectangular area that must be studied. The rectangle method involves placing the transmitting electrodes C_1 and C_2 well outside the rectangle and measuring the potential difference between P_1 and P_2 with a small bipole parallel to C_1C_2 within the area surveyed, to convert it later into apparent resistivity.

The depth of investigation is not well defined with such an array. We should recall that the array will be quite sensitive to surface heterogeneities, but the resistivity will also depend on an average of the deeper resistivity values: we should notice that this is not too dissimilar from a Schlumberger array.

In archeology, when we are looking for buried walls or ditches, this method may be useful due to how it allows us to avoid constantly moving the cables compared with the pole–pole array. On the other hand, the response depends on the orientation C_1-C_2. Thus, it may be useful to create two maps with perpendicular C_1C_2. The apparent resistivities for C_1C_2 in a direction (let us say NS) may be combined with the other direction (EW) based on a constant quantity, an apparent resistivity as a geometric average, namely:

$$\rho_a = \sqrt{\rho_a^{NS}\rho_a^{EW}} \ .$$

Figure 2.32 shows the creation of a rectangle and an actual example related to the map obtained.

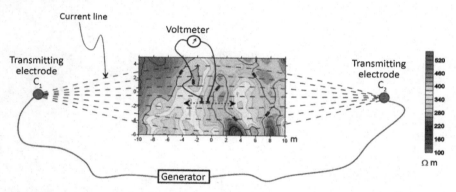

Figure 2.32. *The technique of the resistivity rectangle. We can see that the y direction shows stretched-out areas parallel to Oy. The geometric coefficient is similar to that of a Schlumberger array, but it varies more significantly from one point to the other. For a color version of this figure, see www.iste.co.uk/florsch/geophysics1.zip*

The conductive areas in blue reveal ancient man-made holes dug in the granite base very close to the surface.

2.12. The mise-à-la-masse method

We can use this method when we are trying to characterize a major conductor, buried underground, to which one of the transmitting electrodes is connected. Thus, it is the whole of the conductor that turns into an electrode. Its high conductivity transforms it into an equipotential (as the object is highly conductive, there is no drop in voltage when current is flowing).

Figure 2.33 describes this principle. We are no longer trying to calculate the apparent resistivity, but only to plot the surface potential in the area where we expect to find the conductor.

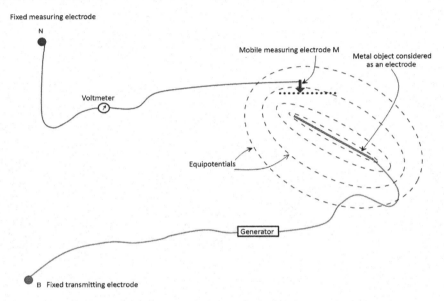

Figure 2.33. *Mise-à-la-masse. By plugging a transmitting cable directly to a conductor, we attempt to follow the conductor because of the equipotential it represents. For a color version of this figure, see www.iste.co.uk/florsch/geophysics1.zip*

Initially, this method was used to contour some sulfide deposits, such as pyrite or galena deposits. These two minerals in massive form have widely variable resistivity values, which are often much smaller than 1 Ωm. If nowadays the mise- à-la-masse method is rarely used for these deposits (mass deposits have been exhausted), it is still very useful if we want to find a conductor (not covered by an insulated bed!) in the subsoil, for example if we want to follow a metal pipe (such as a former water catchment system).

We need a significant resistivity contrast to actually "follow" a conductor enclosed by a resisting element. This is due to the very properties of the laws of electricity. To understand this point, let us imagine a good conductor – let us say with resistivity $\rho_{wire} = 1\Omega$ m – shaped like a cylindrical wire with a radius of a = 1 cm, very long and laid on the ground, barely under the surface of the half-space. Let us assign to the enclosure an average resistivity $\rho_0 = 100\ \Omega\,m$. The resistivity contrast is equal to 100.

Let us conduct this thought experiment represented in Figure 2.34. Let us compare the resistance of a section of the cable, whose length is dr, at the distance r from the center where we will link the conductor to the generator, with the resistance of the soil, at the same distance and with the same center, similarly moving from r to r + dr.

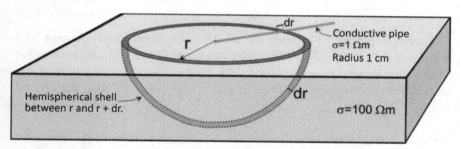

Figure 2.34. *We compare the resistance of 1 cm of pipe at $\rho_{fil} = 1\ \Omega m$ with that of a shell 1 cm thick having resistivity $\rho_0 = 100\ \Omega m$. The laws of electricity verify some geometric laws (which ensure the conservation of the current flow!). There is a surprise in store at the end of the pipe: the pipe is virtually invisible, as is shown in the text. For a color version of this figure, see www.iste.co.uk/florsch/ geophysics1.zip*

The resistance of the wire is[36]:

$$R_{wire} = \rho_{wire}\frac{dr}{\pi a^2} = 1 \bullet \frac{dr}{\pi\left(10^{-2}\right)^2} = 10,000\frac{dr}{\pi}.$$

---

36 Here, as was the case for the ½ space, we use the simple formula for the resistance of a wire.

For the half-space, we consider a shell between the radius r and the radius dr. The corresponding surface is that of the hemisphere: $S = 2\pi r^2$. The resistance of this shell, which separates the hemisphere of radius r from the hemisphere of radius r + dr, is[37]:

$$R_0 = \rho_0 \frac{dr}{2\pi r^2}.$$

Let us carry out the same calculations at a given distance of 5 m from the center, r = 5; in numbers:

$$R_0 = 100 \frac{dr}{2\pi 25} = 2 \frac{dr}{\pi}.$$

Despite being only 5 m from the transmission point, the resistance of a shell of radius r is already 5,000 times weaker than the resistance of a pipe of the same length. In other words, the pipe *carries* the potential of the source from the distance r to the distance r + dr less well than the ground itself.

Some geophysicists have tried to use the mise-à-la-masse approach with water-filled karst conduits; others have attempted to use a trunk and its roots as conductors for a mise-à-la-masse approach with the aim of studying the extent of the root system: we soon face limits that are illustrated by our thought experiment.

Based on the previous calculations, however, we can evaluate which pipe diameter/pipe resistivity ratio will allow us to follow the pipe with the mise-à-la-masse approach.

We always have to compare $R_{wire} = \rho_{wire} \dfrac{dr}{\pi a^2}$ with $R_0 = \rho_0 \dfrac{dr}{2\pi r^2}$.

---

37 The formula that can be used for wires applies here, due to the symmetry of the problem. We can also notice that we would flatten the shell into a cylindrical, if thin, crepe shape.

They can be compared when $R_{wire} = R_0$, namely, after simplifying:
$\dfrac{\rho_{wire}}{\rho_0} = \dfrac{1}{2}\left(\dfrac{a}{r}\right)^2$.

For "detectability", we need to express this equation as an inequality. If the condition:

$$\dfrac{\rho_{wire}}{\rho_0} \leq \dfrac{1}{2}\left(\dfrac{a}{r}\right)^2$$

is verified, we can rely on the mise-à-la-masse approach.

Nevertheless, these calculations are only roughly predictive, as we would need to consider only the portion of the ground near the pipe. Besides, they consider the pipe and the homogeneous ground separately, whereas these two elements should be made to interact. Finally, we obtain orders of magnitude. To go further, we need an actual modeling approach. This is not easy, due to the huge resistivity contrast itself. We could try to consider the pipe as an equipotential, but then we would disregard its resistance, and we would go around in circles.

The formula can be applied to a hollow pipe, but we need to replace its section πa^2 with its actual section. For a hollow pipe made of materials of resistivity ρ_{metal}, we replace πa^2 with $\pi\left(a_{ext}^2 - a_{int}^2\right)$. We can also work by comparing resistance values with units of length, which is possibly a more natural unit for a buried cable or a pipe. Following the same line of reasoning, the resistance per unit length of the half-shell is:

$$\dfrac{R_0}{\text{for 1 m}} = \rho_0 \dfrac{1}{2\pi r^2} \text{ to be compared to } \rho_{wire}\dfrac{1}{\pi a^2} \text{ or directly to the linear}$$

resistance in Ω / m.

2.12.1. *Example*

We can manipulate this equation as we wish. For example: how far can we follow an iron metal pipe whose inner and outer diameter are 3 and 3.5 cm respectively within an enclosure of 300 Ω m? The resistivity of iron is: 10^{-7} Ω m.

According to the equation set up here, we should obtain:

$$r \leq \sqrt{\frac{\rho_0}{\rho_{iron}}\left(a_{ext}^2 - a_{int}^2\right)} = \sqrt{\frac{100}{10^{-7}}\left(0.035^2 - 0.03^2\right)} = 570 \text{ m.}$$

The mise-a-la-masse method comes in handy if we want to follow pipes over large distances, provided that the pipes are very conductive.

Let us imagine that we want to follow a karst conduit of natural water at 30 Ωm, whose radius is 1 m, in a fairly resistant limestone enclosure at 1,000 Ωm. After the calculations, we obtain 5 m, which is not an encouraging result.

Conclusion: if we want to see far (literally), we should use the mise-à-la-masse approach only when dealing with metal conductors. However, this method will be useful for very conductive and confined deposits, such as a localized pollution.

2.13. Time lapse (or simply "monitoring") in electrical methods

All methods lend themselves to time lapse, which allows us to see changes over time. Let us consider two data sets "before" and "after" obtained strictly in the same conditions and separated by an interval T. With a panel array or an electrical sounding, where we systematically use an inversion, we can choose between two approaches:

– separately inverting the measurements "before" and "after", and then comparing the inverted results (generally by difference);

– calculating a difference in the data, and then inverting the difference.

The two methods are mathematically similar as long as the differences are relatively small. However, the inversion algorithms may yield solutions that vary more than required. What characterizes these solutions is that they belong to the same "equivalence pool", i.e. they yield quasi-identical data even for possibly different solutions. This hinders the time lapse. Thus, we need to ensure that the inversion conditions (details and choice in the algorithm) are strictly identical.

2.14. A note about measuring ground resistance

It is often useful to measure an earth connection (of an electrode while prospecting).

With any kind of array (C_1, C_2, P_1, P_2), we only need to connect C_1 and P_1 (linked to each other) to the earth connection that we want to measure, and keep C_2 and P_2 at a distance and far from each other – as was the case for pole–pole arrays. By "far" we mean that the distance between the electrodes is much greater than the size of the electrodes (let us say, at least 20 times). We can carry out certain calculations, i.e. the so-called "62% method"[38] , but the ever-present heterogeneity of the ground makes this rule rather irrelevant (ultimately, it is a protocol rule, so…).

---

38 See, for example, http://www.baselinesystems.com/mediafiles/pdf/62_percent_test_instructions.pdf.

The Spontaneous Polarization Method

3.1. The principle of SP

Let us consider a voltmeter with high input impedance[1] and connect it to the ground with two conductive electrodes: we can observe a potential, which varies according to the position and distance between the electrodes, the weather conditions, the nature of the electrodes, etc.

We have just measured a spontaneous polarization (SP). This is a significant catch-all term, as we will see. At least, this experiment confirms that something happens in the ground, resulting in the production of the potentials[2]. In any case, this is the cheapest applied geophysical method, which everyone can afford!

---

1 The digital voltmeters on the market have high input impedance. This is not always the case for older voltmeters, both those with a needle dial and the cheap digital ones. They should not be used to measure potential differences in the field, since the mere fact of connecting them to electrodes planted in the ground modifies the tension to be measured, in relation to the resistances due to the contact with the ground. Old geophysical apparatuses worked by potential annealing: the operator would set an internal tension against the tension to be measured until they cancelled each other out, resulting in the cancelled measurement of current. Consequently, no current flowed to the device, whose input impedance was therefore infinite. However, this method could only work by transmitting a small current to the ground. Thus, the measurement was not completely neutral for the environment measured. To test the input resistance of a voltmeter, it is enough to use an ohmmeter (another multimeter).

2 Let us recall that this is a misnomer: we can only measure potential differences, and not even that, as in order to do this we need to measure an accumulation of electrons in one of the FETs(#) of the voltmeter. We can only measure potential differences with the currents (which are naturally very weak) applied to a specialized electrical circuit. Even the most resistant electrometer(#) needs flowing charges to take a measurement.

This measurement may involve mapping (or profiles, but mapping is ultimately always richer)[3]. Let us illustrate how this can be done in Figure 3.1.

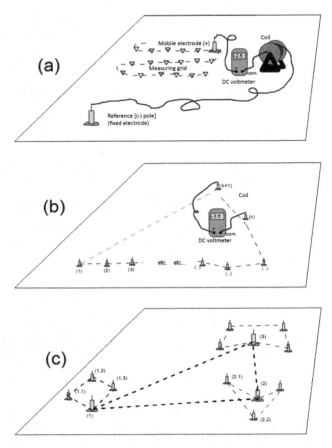

Figure 3.1. *SP protocols. a) Basic protocol, with a reference; b) differential protocol; c) mixed protocol. For b) and c), we should close the system so that we can check the data by returning to the bases (an error anywhere would affect all the following measurements). This is useless in case a), where a point, if it is incorrect, remains isolated as such. Thus, every scenario is possible, provided we can retrieve the potential from any point in relation to the base (often the first point of the campaign)*

---

3 Someone is painting our portrait, representing our face with only a line 2 or 3 mm wide or, in other words, a profile (in the geophysical sense). Our mother is unlikely to recognize us. Whereas the 2D image is a photo of your face. Recognizing anomalies in geophysics is somewhat similar. An actual 2D map is infinitely richer than a mere profile.

This figure presents different mapping approaches, according to the wire length. As our goal is always to create potential differences, we can use a fixed and a mobile electrode or use a step-by-step approach by measuring differences $\Delta V_j = V_{j+1} - V_j$. Thus, the potential at the point (k) in relation to

the point (1) is equal to: $V_k - V_1 = \sum_{j=1}^{k-1}(V_{j+1} - V_j) = \sum_{j=1}^{k-1}\Delta V_j$.[4]

We can proceed by patch and zone, but we should make no mistakes when establishing a reference point. We always start from the principle that the potential in a given point is the sum of the potential differences of all the previous points, in the measuring order, with the (–) pole of the voltmeter always behind.

3.2. The origin of the potentials in SP

3.2.1. *The electrodes*

Without any special precautions (like using so-called "non-polarizing" electrodes, described further on), the SP of the electrodes themselves often constitutes the most significant contribution to the potential observed. We should not consider a voltmeter without its electrodes: the instrument yields the tension at its terminals rather than that of the medium where the electrodes are planted. To prove this, we only need to take a water tray and test different electrode pairs: we always obtain a few dozen millivolt, but the value is much higher when the electrodes are made of different metals (several hundred millivolts).

We only provide here the "voltaic cell" experiment, whose latest developments are known.

Nevertheless, we understand straight away that a problem arises in relation to the measurement of the potentials that may be generated underground... if the potential of the electrodes can vary so widely, how can we separate it from a signal coming from deep below the surface?

---

4 Let us recall that mathematicians are lazy! Instead of writing a sum $S = a_1 + a_2 + a_3 + a_{4+} + \ldots\ldots a_{N-2} + a_{N-1} + a_N$, they write $S = \sum_{k=1}^{N} a_k$, which should be read as "the sum, as k goes from 1 to N, of a_k ". This is very useful when N is a very large number!

It is useful to represent the tension chain that we may encounter in our experiment, which is illustrated in Figure 3.2. Readers can easily conduct this experiment themselves. Plunge two electrodes (for example an iron and a copper electrode) into a (plastic or glass) bowl filled with water. When we connect the voltmeter to the electrodes, we note that the potential is quite significant. We can place a LR6 battery for a few moments at the bottom of the basin, which constitutes a source of tension (and current). We observe then a variation in the tension measured, which will also depend on the orientation of the battery. This small experiment simulates a measurement in the field (clearly, with very different conditions). It evidently shows that measuring SP involves, on the one hand, the potential specific to the electrodes and, on the other hand, a potential generated in the subsoil.

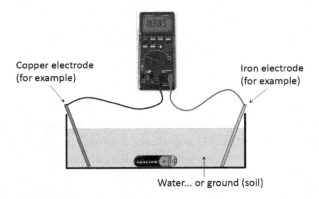

Copper electrode (for example)

Iron electrode (for example)

Water... or ground (soil)

Figure 3.2. *A scale model of the measuring chain for spontaneous polarization. The voltmeter tension is the sum of the tensions generated at the electrode interfaces (which constitutes in itself the first voltaic cell phenomenon) plus the tension generated in the medium, here an artificial redox phenomenon (batteries found on the market)*

As we are interested only in the latter type of potential, we need to find a way of reducing the potential associated with the electrodes or, more precisely, with the interfaces between the electrode and the electrolyte in the medium. The solution is to use two so-called non-polarizing electrodes.

Note that the theory that focuses on electrode potentials is the same as that involving electric cells. The physical law at work here is the Nernst law(#) and the theory is electrochemistry(#). Its importance in everyday life results in a huge amount of material online.

3.2.1.1. *Non-polarizing electrodes*

This is a curious oxymoron, as non-polarizing electrodes cannot exist. What we mean by that is that the electrodes have constant polarization. However, this is not the case yet.

In electrochemistry, there are potential differences: at the interfaces between metal and electrolyte (these differences are those that disrupt the measurement in the first place). The potential depends on the metal–electrolyte pair and the concentration of the electrolyte:

– between two mediums with the same electrolyte but different concentrations;

– between two mediums with electrolytes of different nature;

– between two mediums with different temperatures.

We can create a non-polarizing electrode (more precisely, with constant polarization, even if we accept this term) by plunging a metal in a solution saturated with its own salt. As a result, there can be exchanges between the metal and the solution without any variations in the system state. We still have the same metal in the same electrolyte. The reaction between the metal M and its electrolyte is $M \rightleftharpoons M^{n+} + ne^-$. On the other hand, as the solution is saturated (with excess undissolved salt), we can ensure that the concentration is constant. Therefore, the polarization will be constant[5].

In a SP experiment, we ensure that two non-polarizing electrodes of the same nature are used. Thus, by difference, we can eliminate constant polarization. Otherwise, there will be a discrepancy and an offset.

There are several ways of creating non-polarizing electrodes. All of them must involve:

---

5 It is impossible to measure electrode polarization "on its own", as we cannot measure a potential... By convention, we need to use a universal reference electrode against which we measure the voltage of a given redox pair. For example, this tension is equal to 0.34 V for the pair (Cu, Cu^{++}). The modern reference electrode uses hydrogen gas, see for example https://en.wikipedia.org/wiki/Reduction_potential and the related links.

– the partial immersion of a metal in a solution saturated with the salt of the same metal (the tip being attached to the voltmeter);

– an area that enables the exchange of ions between the electrolyte and the ground. This exchange must be such that the potential is appropriately exchanged. Therefore, there must be an electrostatic effect between the two solutions (inside and outside the electrode) without any massive exchange of these ions (by osmosis). Porous ceramic or wood is used.

Figure 3.3 shows two ways of carrying out this process. The copper/ copper sulfate pair is the one most commonly used, given its low price and how easily we can find these two products.

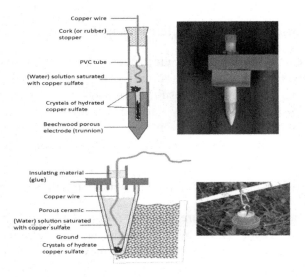

Figure 3.3. *Two ways of creating the electrodes Cu/CuSO₄. The ceramic electrode is made up of a porous candle of the kind that is screwed on bottles to slowly water flowerpots. For a color version of this figure, see www.iste.co.uk/florsch/geophysics1.zip*

3.2.2. *The sources of SP related to the Nernst equation*

The Nernst equation(#) is a thermodynamic equation that expresses electrical potentials in systems where differences are present (regardless of the kind of difference: nature of the ions present, concentrations, temperatures, etc.). In particular, it yields the electrical potentials associated with the redox reactions(#) or "redox" reactions. As we are dealing with redox reactions, we refer to electrochemical potential. Let us point out that section 3.2.1, dedicated to electrodes, is already based on the Nernst equation.

3.2.2.1. *The redox case*

Like the question of the electrodes, several sources of SP involve redox potentials generated in the ground. They are directly linked to the Nernst law, which can be written for a given redox pair as:

$$V_{redox} = V_0 + \frac{RT}{nF} \log_e \frac{[Ox]}{[Red]},$$

where:

– T is the temperature in kelvin;

– R is the ideal gas constant ($8.314 \ JK^{-1} \ mol^{-1}$);

– F is the Faraday constant (one Farad = 96,485 Coulomb);

– V_0 is the standard potential of the redox pair;

– n is the valence of the couple;

– [Ox] is the concentration of oxidizing elements and [Red] the concentration of reducing elements.

3.2.2.2. *The diffusion case*

Concentration gradients in the medium, which enable a differential migration of the ions of a salt (for example Na^+, Cl^-, cations and anions), will also produce a "SP" contribution resulting from the diffusion in the form of:

$$\Delta V_{diff} = \frac{RT(\mu_a - \mu_c)}{nF} \log_e \frac{[C_1]}{[C_2]},$$

where μ_a and μ_c are the mobilities(#) of the anions and cations, respectively, while C_1 and C_2 are the concentrations. The symbol Δ reminds us that we express the potential difference due to the difference of a parameter, in this case concentration.

3.2.2.3. *The temperature case*

Regardless of the application of the Nernst law, temperature appears in the numerator. Consequently, two identical mediums that differ only in temperature will show a potential difference. For example, in the first case, all other things being equal, there will be a contribution:

$$\Delta V_{temp} = \frac{R(T_{hot} - T_{cold})}{nF} \log_e \frac{[Ox]}{[Red]}$$

which will be added.

The cases we have just described can be easily tested. The first by plunging two different metals into a water basin filled with water, the second by measuring the potential where there are two different concentrations (this can easily be done in a water bowl filled with water, by gravity with a more and less concentrated salt) and the last by changing one of the temperatures in the previous example.

In the field, the first case is the most significant. It is especially used to map some polluted areas (household waste, heaps of metal). This situation leads us to what we call "geobatteries", with strong redox potentials. We can find an example in Naudet *et al.*[6], which is illustrated in Figure 3.4.

---

6 https://hal.archives-ouvertes.fr/file/index/docid/330848/filename/hess-8-8-2004.pdf.

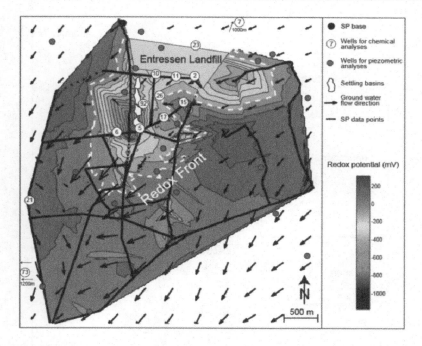

Figure 3.4. *Map of the redox potential on the site of the Entressens landfill. The areas with marked anomalies denote significant chemical activity (from Naudet et al.). For a color version of this figure, see www.iste.co.uk/florsch/geophysics1.zip*

A strong SP signal is associated with the active concentrations of the landfill.

However, the most important application of "redox" SP involves the signal created by sulfide deposits, which have often been called "geobatteries", as illustrated in Figure 3.5.

Marked and always negative SP anomalies are detected directly below sulfide deposits. In 1960, Sato and Mooney put forward a model that involved different redox potentials (or oxidation states) and depths. It included a sort of vertical battery with the (+) pole facing downwards. This model has been discussed as it assumed that the deposit was intercepted by a water table, whereas "water table free" deposits produced similar anomalies. However, it is estimated that the oxygen content, which decreases with depth, could be enough to generate such anomalies.

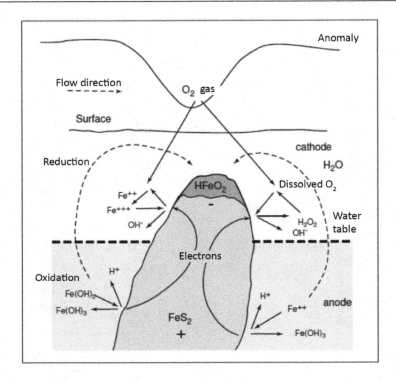

Figure 3.5. *Sato and Mooney's model of geobattery SP for sulfide deposits*

This type of SP related to metal deposits is handy when we are looking for buried metallic objects, naturally, provided that they are not isolated from ground. Figure 3.6 shows a sample anomaly on a vertical borehole metallic casing.

3.2.3. The sources of SP related to the electrofiltration equation and the application to aquifers and leakages

Let us take a cylinder filled with sand and open at both ends (with fine gratings). Let us place some measuring electrodes on each side (for example the gratings may be used as electrodes) and let us force a pressurized water stream to go through the sand. We can observe a potential difference between these electrodes, with a positive potential on the downstream side (Figure 3.7).

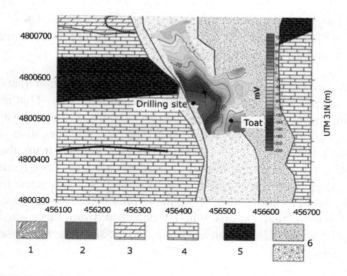

Figure 3.6. *A SP anomaly on a corroded vertical casing, at the Cabrespine site (drawn from Galibert, Florsch and Camerlynck, personal communication and work with the students). Geology: (1) axial zone (quartzites and schists), (2) limestone slabs d1-2, (3) dolomites d2-3a, (4) white limestone d3a, (5) calcshists d3b-7 and (6) alluvial terraces. For a color version of this figure, see www.iste.co.uk/florsch/geophysics1.zip*

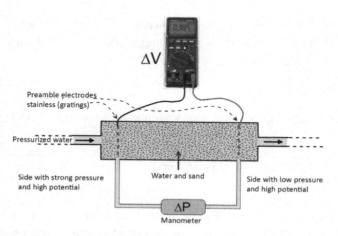

Figure 3.7. *An electrofiltration experiment that shows a pressure gradient associated with a potential gradient. This remark, which is now backed up theoretically, gives us the possibility of defining water flows in the subsoil. This method, which is still at the center of a substantial amount of research, is routinely employed for several purposes, for example to monitor leakages in dams*

The theory governing this process is quite complex. Those adept at equations will find something to work with on Andre Revil's website at: http://www. andre-revil.com/publi.html.

We can, however, provide a summary: it turns out that (sand) grains present a surface layer and coating with a negative charge. This is the case for most minerals. The coating generates an "electrical double layer"(#) that results in cations (positive ions) attaching to this negative layer by electrostatic attraction. However, when we get farther from the surface (around 10 nm as a reference point), the cations are less strongly attached to the surface negative charges, and moreover they are subjected to thermal motion like all other molecules. If we make pressurized water flow, it is due to mere viscosity that cations will be detached from these surfaces and dragged downstream. The result is the equivalent of a battery whose (+) pole is downstream.

If we note pressure as P and the potential always as V, we find experimentally that:

$$V_{downstream} - V_{upstream} = C\left(P_{upstream} - P_{downstream}\right),$$

where C is a so-called "coupling" constant.

We should note that $P_{upstream} > P_{downstream}$ and that $V_{downstream} > V_{upstream}$. However, physicists are used to employing gradients, so that they will write this equation as (equivalent):

$$\overrightarrow{\nabla V} = -C\overrightarrow{\nabla P}.$$

The coefficient C in the aforementioned experiment can be easily measured.

Theoreticians link C to physical parameters:

$$C = \frac{\varepsilon \zeta}{\sigma \eta},$$

where ε is the electrical permittivity of the fluid (dielectric constant), ζ is the "zeta potential", which corresponds to the potential drop when we get farther from the electrical double layer, σ is the electrical conductivity of water and η is its dynamic viscosity. This is the so-called Helmholtz–Smoluchowski's equation(#).

Let us use this information in the field: there must be a relationship between potential and water flow. In other words, SP should tell us something about the water tables.

Maurice Aubert, a Frenchman, worked within this framework for a water company, whose name will not be revealed, in Auvergne close to a town called Volvic. The hydrological context is peculiar. There is an impermeable crystalline basement over which "flow" the aquifer, which is fairly thin (except for the hollows in the crystalline basement) in the return flows, volcanic ashes and permeable bullous lava. To use his notation[7], let us express the depth of the aquifer in the point with coordinates (x, y) as $E(x,y)$, this depth in a point on a reference site for SP[8] as E_0 and the potential in the same point (x, y) as $V(x, y)$. Aubert pointed out that we can write a linear relation between the potential and the depth of the aquifer as:

$$V(x,y) = K\left[E(x,y) - E_0\right].$$

Let us quote here Maurice Aubert, who invented the notion of "SSP" (Surface SP):

"The value of K varies according to the nature of the ground, ranging in our experience from -1 mV/m for altered ancient types of ground to -7 mV/m for recent types of ground. In this model, the SSP is the zero equipotential surface, as the other equipotentials can be deduced from the SSP by translation. The minimum potential lines are situated directly below the thicker lines of the volcanic capping (talwegs), whereas the maximum potential lines correspond to the watershed lines. These lines are generally left mostly unchanged by a variation of K, whereas their slopes depend directly on E°. This model is said to be empirical as it does not rely on any specific physical process"[9].

---

7 Here is a link to a summary document about this topic: https://www.erudit.org/fr/revues/rseau/2003-v16-n2-rseau3313/705505ar/. The document is in French, but it contains some references in English.
8 Thus, we should place the SP reference, linked to the (−) terminal of the voltmeter, somewhere where depth is known (well, drilling, outcrop of the water table, etc.).
9 *"No specific physical process"* ... This may have been the case when Aubert published this book, but several authors have related this empirical law to the theory of electrofilitration since then.

As a result, the depth of the aquifer in relation to the surface can be deduced with this equation (from now on, we will omit the reference to the point (x, y) when we write these equations). Therefore, we obtain:

$$E = E_0 + \frac{V}{K}.$$

Aubert wrote it by including altitudes– that of the top of the water table H(x, y) and that of the topographic surface h(x, y) – in the following form (while taking into consideration that E = h – H):

$$H = h - \frac{V}{K} - E_0.$$

The case presented by this author involves an aquifer area with volcanic ground (Figure 3.8).

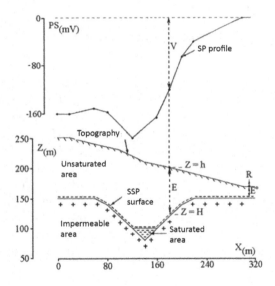

Figure 3.8. *Maurice Aubert's definition of SSP surface*

Afterwards, this equation has been generalized, as a theoretical basis was included and K was linked to the coupling coefficient. However, regardless of these steps forward, the coefficient K will always have to be determined, for example by finding another site where E is known, let us say in the point M. Therefore, we obtain:

$$K = \frac{E(M) - E_0}{V(M)} .$$

We should be careful not to rely too much on these laws when defining an aquifer. The electrofiltration SP is never alone, and clay or other sources of potentials may thwart these approaches altogether.

Let us consider the example of the La Proutière[10] site in Vendée. The context is crystalline, with altered gneiss close to the surface. The geological nature of the ground suggests a Wyns-like model, which has already been mentioned.

If the topographic surface seems regular, the subsurface is marked by strong contrasts, where we can find very altered areas (sandy soil) laterally followed by parts of intact gneiss. In the work quoted (which can be easily downloaded), it has been possible to correlate the SSP surface with the (underground) topography of the water table, with a correlation quality that geophysicists deem satisfactory.

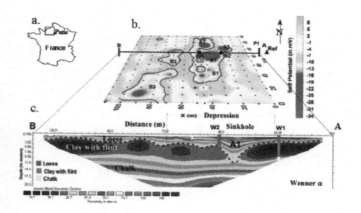

Figure 3.9. *SP anomaly (map) and electrical tomography of a sinkhole area in Normandy[11]. SP marks the infiltration zones by means of negative values compatible with water flowing downwards. For a color version of this figure, see www.iste.co.uk/florsch/geophysics1.zip*

---

10 https://hal.archives-ouvertes.fr/hal-01548708.
11 Drawn from "Least squares inversion of self-potential (SP) data and application to the shallow flow of ground water in sinkholes" GRL, vol. 33, L19306, doi:10.1029/2006 GL027458, 2006 by Jardani, Revil, Akoa, Schmutz, Florsch and Dupont. This document can be read from http://www.andre-revil.com/publi/html.

Another application of SP involves detecting sinkholes(#) that cannot be seen above ground yet, or more generally vertical leaks in aquifers. Thus, the anomalies are negative, as it is weight that carries the water downwards: the surface is upstream – water flow. Figure 3.9 illustrates a good example.

The Induced Polarization (IP) Method

4.1. The principle of induced polarization

Works carried out around 1913 and mentioned in Chapter VIII (3 pages) of the aforementioned landmark book by Conrad Schlumberger result in induced polarization ("IP"). The observation is simple: right after a direct current has been transmitted to the subsoil (with electrodes C_1C_2 — we can imagine a Wenner array or a…Schlumberger array), the potential measured at the measuring electrodes P_1P_2 does not immediately drop back to zero, decreasing instead in a roughly exponential way.

Consequently, we are led to think that there are elements in the subsoil itself that can become electrically charged and then restore this charge when no current is flowing. The decrease may range from rapid (let us say some milliseconds, the shortest periods involving purely electromagnetic phenomena)[1] to slow, occurring over several seconds. The range is even wider, but in practice we work with this timescale.

If the precise mechanisms had still to be identified in Marcel and Conrad Schlumberger's times, today we know what leads to the reversible storage of charges in the subsoil. There are several concurrent phenomena:

– The "membrane polarization"(#) linked to the difference in mobility among the ions (moving speed) in an electrical field and associated with short relaxation times. It concerns all of the medium.

---

1 Technically speaking, *everything is electromagnetic*, but we are used to distinguishing what involves induction ("Foucault currents") or propagation (radio) from what results from "electrochemical" phenomena (such as electrolysis or battery effects).

– The "Maxwell–Wagner polarization"(#) involves an accumulation of charge on both sides of the interfaces that separate mediums of different conductivity and/or electrical permittivity.

– The polarization of the "electrical double layer"(#) occurs especially near grains (for example quartz) whose surfaces are (negatively) electrified. These surfaces include a layer of "counter-ions", cations, that are locally involved in electroneutrality. If we move a few nanometers away from the surface, these cations are less restrained than those nearest the surface. The whole cloud of positive ions attached more or less strongly to the particle is deformed by the action of the external electrical field imposed by the generator. It is this deformation that determines the polarization.

However, the most intense phenomenon of polarization, by far, takes place in the presence of particles with electronic conductivity, namely when it is electrons and not ions that are charge carriers (as is the case for metals). This is true for magnetite, pyrite, galenite, graphite, etc. These particles are several orders of magnitude more conductive than the average background. After the switching on, the particles become equipotential (no difference in tension inside the particle). Therefore, the charges (electrons) are distributed in the particle to compensate exactly for the external field such as it would exist in relation to the particle if the particles were not there. In doing so, the particle becomes a small electric dipole, which will somehow in turn attract and repulse (according to their sign) ions taken in the surrounding areas, as well as determine polarization of the whole. We should point out that IP will be also useful if we want to recover buried macroscopic metallic objects, provided that the metal is not coated with insulating material (household waste such as the frames of washing machines, etc.).

As the last mechanism, called "electrode polarization", is the most intense and it is hard to tell apart the other mechanisms, which are still being researched, we mention here only aspects related to electrode polarization. The polarization mechanism on a particle level is illustrated in Figure 4.1.

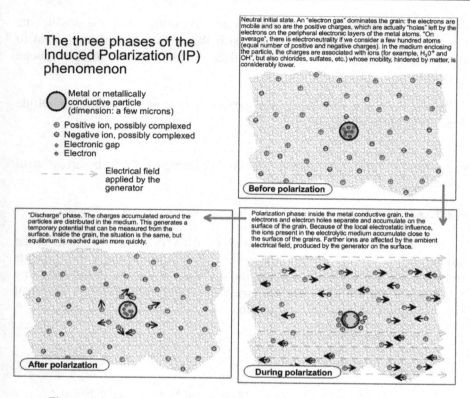

The three phases of the Induced Polarization (IP) phenomenon

Metal or metallically conductive particle (dimension: a few microns)

⊕ Positive ion, possibly complexed
⊖ Negative ion, possibly complexed
• Electronic gap
• Electron

– – – – – ⇒ Electrical field applied by the generator

Neutral initial state. An "electron gas" dominates the grain: the electrons are mobile and so are the positive charges, which are actually "holes" left by the electrons on the peripheral electronic layers of the metal atoms. "On average", there is electroneutrality if we consider a few hundred atoms (equal number of positive and negative charges). In the medium enclosing the particle, the charges are associated with ions (for example, H_3O^+ and OH^-, but also chlorides, sulfates, etc.) whose mobility, hindered by matter, is considerably lower.

Before polarization

"Discharge" phase. The charges accumulated around the particles are distributed in the medium. This generates a temporary potential that can be measured from the surface. Inside the grain, the situation is the same, but equilibrium is reached again more quickly.

After polarization

Polarization phase: inside the metal conductive grain, the electrons and electron holes separate and accumulate on the surface of the grain. Because of the local electrostatic influence, the ions present in the electrolytic medium accumulate close to the surface of the grains. Farther ions are affected by the ambient electrical field, produced by the generator on the surface.

During polarization

Figure 4.1. *The (microscopic) mechanism of electrode polarization. For a color version of this figure, see www.iste.co.uk/florsch/geophysics1.zip*

On a macroscopic scale and when several electronically conductive particles are present, the individual effects add up. It is the return to equilibrium of this group of dipoles created around the particles that we observe with two measuring electrodes located on the surface.

On a theoretical level, the elements that enable the IP signal are the same as those enabling SP. The underlying theory involves the laws of electricity and statistical physics. The former determine the attractions among electric charges, whereas the latter shows how thermal motion and Brownian motion undo the work carried out by the electrostatic forces. All of this can be modeled by coupling the "Nernst–Planck equations"(#) with "Poisson's equation"(#), creating what is called the "Poisson–Nernst–

Planck"[2] system or "PNP" (nothing to do with transistors). To confirm its universality, let us mention that it is also the system of equations used to study the transmission of nerve impulses or calculate our mobile phone batteries.

IP has been and still is the most powerful way of characterizing sulfide deposits of economic interest.

4.2. Three types of measurements: temporal, frequency and spectral

4.2.1. Temporal IP

This is the traditional method, which is still very commonly used in the mining industry. The operator transmits a direct current for a few seconds and then cuts the power, observing the decrease in tension at the receiving electrodes. Naturally, this can be done when carrying out classic electrical prospecting, of which IP is an extension.

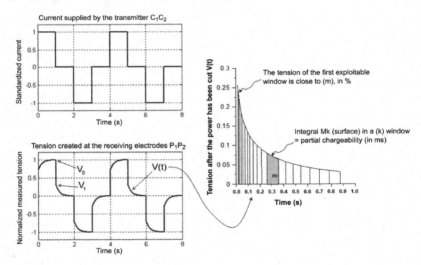

Figure 4.2. *Temporal induced polarization (TIP) signal. The IP signal is deduced from the decay curve of the tension: integral and/or characteristic time constant*

2 For those very interested in the mathematical aspect, a scientific article is available that provides an overview: https://www.ncbi.nlm.nih.gov/pmc/articles/PMC3122111/pdf/JCPSA6-000134-1941011.pdf.

In practice, the signal measured is noticeably weaker than that associated with the transmission. It represents a small percentage of the signal measured as the current is transmitted. Moreover, it adds to the SP of the electrodes and to natural SP. Thus, we prefer using a signal with alternating transmissions, as Figure 4.2 illustrates. For example, we transmit the current in a given direction for a second, then we cut the power and we observe the decrease for a second, and then we transmit another current for a second in the opposite direction, observing the decrease again for a second. This cycle, here with 4-s intervals, is started several times, and devices calculate the average for the series of decreases, after ordering them all in the same positive direction (we refer to "stacking" data).

Based on the decay curves, now recorded by PCs, we can measure several parameters:

– the value of tension just after the power has been cut (we mean by that we need to wait first for a few milliseconds to avoid "merely" electromagnetic effects). Based on the tension measured just *before* the power is cut, let us say V_0, and the tension measured for example 10 ms later, let us say V_1, we can deduce a "chargeability" designated as (m) and defined by: $m = \dfrac{V_1}{V_0}$, non-dimensional, and expressed in percent (%);

– the integral (therefore, the surface under the curve) in one or several chosen time windows. This is also a chargeability, designated as (M), which may be partial (reduced window between two given times) or total (integral of the whole curve, but this is a misnomer as the observation period, and therefore the integration period, is necessarily restricted to the non-transmission period). If we denote V(t) as the voltage after the power has been cut (together with a short delay, as we have already pointed out) and $M_{t_1}^{t_2}$ as the chargeability for the time window $[t_1, t_2]$, we can obtain this chargeability, which we also standardize as $M_{t_1}^{t_2} = \dfrac{1}{V_0} \int_{t_1}^{t_2} V(t)dt$, expressed in ms, or a standardized chargeability for the measuring period, namely: $\tilde{M}_{t_1}^{t_2} = \dfrac{1}{V_0(t_2 - t_1)} \int_{t_1}^{t_2} V(t)dt$, which is once again dimensionless;

– the time constant (τ), for a simplified decay model like $V(t) = me^{-\frac{t}{\tau}}$. If this model is correct – and if we also discard the very beginning of the signal (some milliseconds after the power has been cut, as it is polluted by a short electromagnetic response) and take into consideration the fact that we measure during the period T/4 after the new transmission – we can approximate (τ) with the following equation: $\tau \simeq \dfrac{1}{V_0} \displaystyle\int_{t=\varepsilon}^{T/4} V(t)dt$.

These integrals are expressed as continuous here. In reality, $V(t)$ is sampled with values V_k provided by devices as representative of the average for the windows $[t_k, t_{k+1}]$, with k going from 1 to N (for example). We should recall that the integral can be interpreted graphically: it is the area under the curve. Consequently, we simply obtain for the last equation that:

$$\tau \cong \frac{1}{V_0} \sum_{k=1}^{N} V_k \left(t_{k+1} - t_k \right).$$

Theoreticians have long acknowledged that there is a relationship between the time constant and the diameter of the grains responsible for the IP signal. This relationship can be expressed as:

$$\tau = C_0 \frac{a^2}{D}$$

where a is the radius of the grain, D is a constant related to the diffusion of ions and C_0 is a constant. This equation is verified and useful for a large number of materials, such as sandy soils, *but it is not suitable for metal particles*, for which the concentration of the electricity also plays a role.

We know something for a fact about chargeability (regardless of its definition) and the *concentration* of metal particles. For a given type of material, there is a good degree of proportionality between these two parameters, but the proportionality constant must be established for a given site and target. This proportional relation allows us to "convert" an IP signal into the concentration of polarizable material.

Figure 4.3 shows a sample IP three-dimensional (3D) tomography.

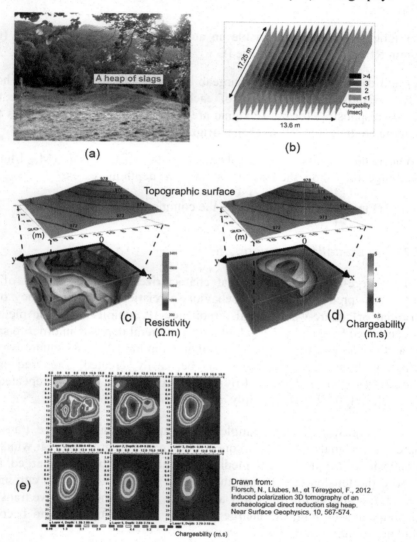

Figure 4.3. *The "target" is a heap of slags produced by an old steel industry in the mountains of Ariège. The image illustrating chargeability (on the right) shows that this parameter is almost independent of resistivity (on the left). This result was obtained by inverting the data acquired because of parallel sections that were sufficiently close to one another for us to refer to 3D tomography (rather than a simple group of 2D sections). For a color version of this figure, see www.iste.co.uk/florsch/ geophysics1.zip*

a) an image of the site with the location of the heap has been obtained geophysically;

b) tight pseudo-sections enable an actual 3D exploitation protocol (the program RES2DINV has been used);

c) and d) Respectively and chargeability, respectively. We can see how the two signals are independent of each other. This is typical when the minerals responsible for the IP signal are electronically conductive; it is the magnetite contained in the slags that originates the signal;

e) horizontal cross-sections obtained with RES3DINV (M. Llubes processing); they show the loss of resolution as depth increases.

The 3D images were created with the commercial software Voxler.

4.2.2. *The so-called frequency IP*

The charge–discharge cycle that characterizes the measurement of IP allows us to predict a different behavior in relation to the frequency of a sinusoidal or square-wave signal. Frequency IP exploits this. The method was conceived by geophysicists looking for mineral deposits in order to save time and use a more flexible approach than temporal IP. As a "square-wave" alternating transmission current can be easily created, the frequency approach has been widely used by mining companies. What is repeated is not a cycle (+I, 0, –I, 0) but simply an alternation (+I, –I).

The "frequency" idea is simple. As Figure 4.4 shows, if we transmit square-wave signals with low frequency (therefore a large interval which is the opposite of frequency), the medium has the time to become charged, and therefore the tension at P_1P_2 has the time to increase. When the current is inverted, it is simply the other way around. If, on the other hand, we transmit high-frequency current, the charge does not have enough time to become complete, so that the amplitude is lower.

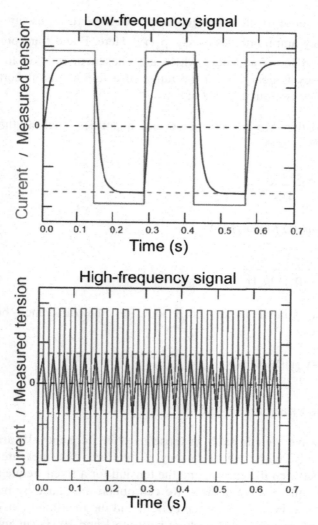

Figure 4.4. *A summary illustration of the frequency method. The amplitude of the resulting signal depends on the frequency and allows us to calculate the frequency effect (FE). For a color version of this figure, see www.iste.co.uk/florsch/ geophysics1.zip*

Low-frequency amplitude is greater than high-frequency amplitude.

Electrically, we measure "RMS values"(#) of currents and tensions. Thus, we distinguish between high-frequency amplitude, expressed as apparent resistivity because of the usual geometric coefficient, let us say $\rho_a(F)$, and low-frequency amplitude, let us say $\rho_a(f)$. Here, F and f are precisely high frequency and low frequency, which are chosen in relation to our target. Low frequency is generally in the range of a few tenths of hertz, whereas high frequency is around 10 Hz or so.

Based on this, we can define several quantities, such as the *frequency effect* (or FE), with:

$$FE = \frac{\rho_a(f) - \rho_a(F)}{\rho_a(f)}$$

but also the *metal factor* (MF):

$$MF = \frac{\rho_a(f) - \rho_a(F)}{\rho_a(f)\rho_a(F)}$$

The term "metal factor" also indicates the difference between the conductivities:

$$MF = \sigma_a(F) - \sigma_a(f).$$

4.2.3. *Phase measurement IP*

Square wave signals such as those used in frequency IP are employed here, but only with one frequency. Therefore, we measure the phase difference between the current and the tension for a fixed frequency of a few hertz. With square wave signals, we can obtain the phase by multiplying a reference signal in phase with the current and the measured tension. We can obtain a good sensitivity by working between 1 and 10 Hz (in any case if we are looking for electronically conductive particles, such as most sulfides).

4.2.4. Spectral induced polarization

For a frequency (f), the generator produces a current of this kind: $i(t) = I_0 \cos(2\pi f\,t)$.

The potential measured is of this kind: $v(t) = AI_0 \cos(2\pi f\,t + \varphi)$.

The functions, namely the amplitude A and the phase φ, depend on the frequency. The measurement involves then scanning the frequencies with at least 3 or 4 points per decade, and then plotting $A(f)$ and $\varphi(f)$.

This is the correct way of measuring IP. It goes through a wide range of frequencies (typically from 0.001 Hz to 20 kHz). For convenience, geophysicists prefer to use complex numbers. This is equivalent to the aforementioned equations. Thus, we can describe a complex current as follows:

$$\tilde{i}(t) = I_0 e^{j2\pi ft}$$

($j^2 = -1$ for electrical engineers, leaving (i) to the current)

and $i(t)$ is the real part of $\tilde{i}(t)$.

Therefore, the tension at $P_1 P_2$ is the temporal function:

$$\tilde{v}(t) = A(f)e^{j\varphi(f)}e^{j2\pi ft} = Ae^{j\varphi}\tilde{i}(t),$$

where the amplitude A and the phase φ are functions of the frequency f.

We can find $v(t)$ by considering the real part of $\tilde{v}(t)$.

The complex numbers allow us to find a simple multiplicative relation between $\tilde{v}(t)$ and $\tilde{i}(t)$. With the same $A(f)$ and $\varphi(f)$ as before, the whole response spectrum, the "transfer function" between the current and the tension, is the complex ratio depending on the frequency f:

$$Z(f) = \frac{U(f)}{I(f)} = A(f)e^{j\varphi(f)}$$

Spectral induced polarization (SIP) ultimately involves plotting these two functions, namely the amplitude A(f) and the phase $\varphi(f)$, which are certainly real functions. This principle is illustrated in Figure 4.5.

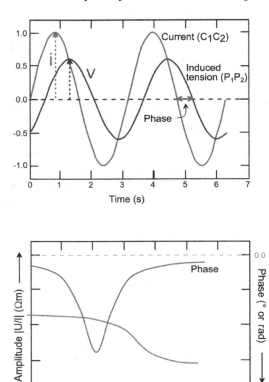

Figure 4.5. *Spectral induced polarization involves studying the complex impedance of the ground in relation to frequency. It consists of exploiting the amplitude and the phase of the transfer function, which links the current (input, electrodes C_1C_2) and the voltage (output, electrodes P_1P_2). For physicists, this is a traditional electrochemical technique. However, its application in the field is not immediately obvious, as the signals are weak and noisy, and the polarization signal of the particles must be separated from several electromagnetic phenomena that are mostly irrelevant for geophysicists. Our colleague Andreas Kemna provides several links on Google Scholar: https://scholar.google.de/citations?user=uoQdyjAAAAAJ&hl=de. Readers will also be able to find numerous links on André Revil's Website (more technical): www.andre-revil.com/. For a color version of this figure, see www.iste.co.uk/florsch/geophysics1.zip.*

The development of SIP is still a challenge in research (as usual, however, readers will be able to find several references online by using the keyword "spectral induced polarization").

These curves are often adjusted with a curve model. The most famous of them is the "Cole-Cole equation"(#) model. Initially designed for permittivity values, it has been extrapolated to consider resistivity or conductivity[3].

---

3 Tarasof and Titov (2013) provide more information: https://academic.oup.com/gji/article/195/1/352/608470/On-the-use-of-the-Cole-Cole-equations-in-spectral#10100707.

Equipment

5.1. Electrodes, wire bobbins and cables

We mentioned non-polarizing electrodes in Chapter 3. They are useful for IP as well, as they cause less (electrode) noise than metal electrodes.

In most cases, we can use any kind of post with a diameter ranging from few millimeters to a few centimeters (according to whether we want to "send" more or less current). They must be solid, as we will most often need a hammer to sink them, ranging from a few centimeters to a few decimeters deep.

The cable should not be permanently connected to the post. These connections can be irreparably damaged by hammer blows. Jump cables of the right size are certainly the best option. On the other hand, a banana plug is connected to the coil (Figure 5.1).

Except for short cables, the bobbin is mostly required to roll and unroll the wires, for example when we are dealing with SP or using arrays, and even for classic arrays such as Wenner or Schlumberger arrays.

Single-strand wires, which must be insulated (not only for safety reasons!), need not have a large section, as the currents used generally do not exceed 1 A. The main requirement is that they are quite solid. It is also better to use a plastic sheath that slides easily on the ground (think about rolling back the wires).

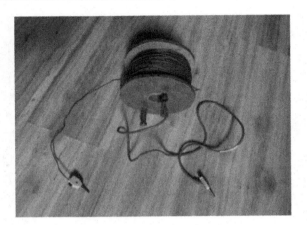

Figure 5.1. *A bobbin used for electrical prospecting*

We should always verify that the cables are in good condition at the beginning of the "campaign". A cable default in the transmission line can be well detected (even if it makes us waste time), whereas in the receiving line this may be less apparent. A cable unconnected to the ground constitutes a good aerial for the electrical field, which receives background radiation in addition to that produced by the transmission line. The input impedance of modern voltmeters is such that we cannot notice anything.

One or even two multimeters, some spare cable, some "electrical tape", connectors, jump cables and "banana plugs" (where we can quickly plug a bare cable end) are indispensable objects that we should have in our electrical tool bag if we want our geophysical campaigns to be successful.

5.2. Equipment and methods used for spontaneous polarization

A long coil (typically more than 100 m), three connecting cables at most (according to the sockets on the coil), two non-polarizing electrodes (with maintenance water and copper sulfide, as well as a Tupperware container to line up the electrodes), and finally a (digital) voltmeter with high input impedance make SP the most economical geophysical method.

It is better to have a voltmeter that displays tenths of mV (direct current mode), even if in the end the precision we work with is only in the range of

1 mV. The tension may be positive or negative, and we will ensure that we correctly identify the reference electrode, in relation to which every tension value will be measured. This is the point zero volt, which must be connected to the (–) terminal of the voltmeter. A tiny slip in this sense may compromise the interpretation of the data.

Even when using digital multimeters, we still need to pay attention to input impedance and verify with an ohmmeter that its value is high.

5.2.1. *Getting rid of disturbances and taking a good SP measurement*

It is nearly impossible to work in rainy weather. The irregular way in which water flows in the ground, together with frequent fingering, interferes with SP through electrofiltration.

Another type of interference is grass with its root layer. While we are trying to measure weak signals, let us say a few millivolt, originating deep underground (by that we mean below the root layer), the potential difference between the base of the roots and the tip of the grass (or between the base and the tip of the roots themselves) may exceed 100 mV[1]. This is clearly a spatially irregular potential, which is therefore irreducible. Moreover, even if there is no grass, the water potential[2] on the surface, linked to the rapid variation in the water content in the first centimeters, produces interfering signals with the potential to obstruct the measurement, particularly due to the marked evaporation (dry weather with a wet subsoil).

Ultimately, taking a good SP measurement *always* requires us to dig a hole. The electrode will be placed in this hole and will touch the area under

---

1 In forestry, the SP measured between two heights on a trunk allows us to follow the sap flow.

2 According to https://en.wikipedia.org/wiki/Water_potential: "Water potential quantifies the tendency of water to move from one area to another due to osmosis, gravity, mechanical pressure or matrix effects such as capillary action". This is not a type of electrical potential. However, it is legitimate to study the relationships between water potential and electrical potential (still, it is electrostatic forces that are at work in both cases), which are still being researched.

the roots. In this case, it is possible to line the bottom of the hole with wet bentonite(#)[3], so that we can ensure there will be good contact (see Figure 2.5).

5.2.2. *Measurement fluctuations*

Once the electrode has been placed, and in relation to the desired sensitivity (from some millivolts to a tenth of a millivolt), the settling time will range from some seconds to several minutes. Moreover, especially when long lines are used, the measurement cannot not be stable, as everything changes over time: the temperature at the electrodes, the evapotranspiration (root pumping), the water flow as well as low-frequency telluric currents.

5.3. Equipment and approaches used for direct current methods

When carrying out mapping (using "arrays") or soundings, we need four electrodes and consequently four coils. Electrical Reistivity Tomography (ERT) (see section 2.7) arrays may be used by hand or with a commutation system. If we employ these types of array, which require a large number of electrodes, a controlled switch makes it easier to acquire the data. This point will be treated in Volume 3 (details in Bibliography). Otherwise, we can easily build a manual switch, as illustrated in Figure 5.2.

Figure 5.2. *Manual switch used with 25 electrodes. On the left, the two cables lead to the transmitter (the red cable) and the receiver (the black cable). The ribbon cable linked to the connector DB25 distributes the current according to the position of the switches. Each strand of the ribbon cable is connected to an electrode by means of jump cables. The on–off button does not supply power, but it connects the two wires with the ribbon cable. It is used mostly as a safety measure*

---

3 We can also use a local type of mud instead of bentonite.

As for the measurement circuit (electrodes P_1P_2), we can generally use metal electrodes made up of stainless steel, copper, etc. Their diameter will range from a few millimeters to a few centimeters, according to the desired depth of investigation (more precisely, according to the ground resistance we wish to reach), and their length will be in the range of a few decimeters, in relation to the desired current.

If the expected signals are weak to very weak, it is still recommended that we use non-polarizing measuring electrodes. However, this is not always possible. The electrodes of ERT arrays may be alternately emitting and receiving. Polarizable electrodes remain...polarized for a few minutes after current has been supplied (a battery effect...). Some non-polarizing electrodes on the market cannot be used, as they irreversibly deteriorate if they are used as emitting electrodes. $Cu/CuSO_4$ electrodes with liquid electrolyte minimize this obstacle and can often be used as receiving and emitting electrodes.

5.3.1. *Adapting the impedance for the transmission of electricity*

Let us present a few of the limiting factors in an electrical prospecting campaign.

– *Factor 1*: The main factor governing the design of a current transmitter is its power P. It determines the size of the device and the current source (cell < battery < generator), its weight and finally the depth of investigation it can reach. However, the power must be correlated with the ground resistance (total, C_1+C_2).

– *Factor 2*: A limiting factor involves the maximum tension U_{max} that may be produced by a generator. This limitation is technological. A first leap occurs when we move from the tensions of low-voltage electronics (15 V) to the tension of household installations (200 V). A second qualitative leap allows us to exceed 3,000 V with devices that can be found on the market. Manipulating electrodes or cables starts becoming dangerous. This should be avoided, and we encourage readers to go through the safety measures again: we cannot protect ourselves from electrocution when carrying out electrical prospecting.

– *Factor 3*: It is related to ground resistance, let us call it R_T. It depends on the size of the electrode and the resistivity of the ground in which it is sunk. When it seems excessive, we use the watering technique with saltwater. This technique, which is very effective, involves an increase in the size of the electrode: in a resistant medium (let us consider a slightly dry sandy soil), it is the post + salty area group that becomes "one" electrode. However, scattering some salt on the ground is not always recommended.

– *Factor 4*: The current has a maximum value I_{max} beyond which the device "is fried". This is an electronic limitation.

Let us consider in more detail the question of adapting the impedance. Thus, let us think about the characteristics of the generator again: a maximum given power P, a U_{max} and a maximum current I_{max}. We know that P = UI, and therefore we can plot U as a function of I. This is a hyperbole, which is represented in Figure 5.3.

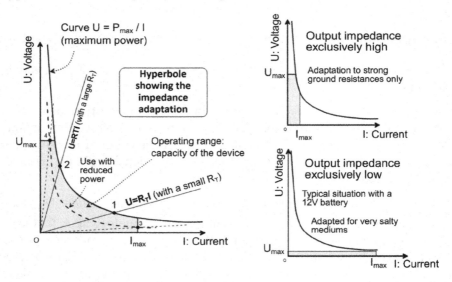

Figure 5.3. *The question of adapting the impedance. We represent Ohm's law, U = RI, with weak ground resistance in case (1) and strong ground resistance in case (2). We can see immediately that in case (1) the generator must supply a strong current for a small tension, whereas in case (2) it must supply a weak current for a strong tension*

This is what adapting the impedance means. Devices (all of them!) work within an operating range, which in practice includes more or less the range of (total) ground resistance between 100 and 5,000 Ω, and are limited by a maximum tension and current.

Some types of ground are unsuitable. In some geologically ancient areas, everything consisting of easily soluble ions was dragged away by the rain a long time ago. Provided that the weather is dry, we will obtain huge ground resistance values, like in case (4). The opposite is also true, for example if we want to determine the thickness of the sand on a marine (and consequently salty) beach, where the resistance values may go down to a few ohms, as is shown in case (3). If the electronic component is not adapted, it will be a disaster for the device[4].

It is interesting to consider the consequence of the adaptation (or inadaptation) of the impedance on the precision of the apparent resistivity measurement. We obtain $\rho_a = K \dfrac{\Delta V}{I}$. The "classic" formula[5] used to calculate relative error leads us to write:

$$\left|\frac{\delta\rho_a}{\rho_a}\right| \le \left|\frac{\delta\Delta V}{\Delta V}\right| + \left|\frac{\delta I}{I}\right|$$

The error related to the current is due mainly to the instrument, and it is generally limited. On the other hand, $\delta\Delta V$ will certainly contribute a term linked to the instrument but also an irreducible term related to "natural noise", i.e. a signal that cannot be controlled in any way (earth noise, potential of the electrodes, etc.). Therefore, the term that we need to try to make as small as possible is $\left|\dfrac{\delta\Delta V}{\Delta V}\right|$. Since the numerator cannot be reduced,

---

4 On a beach on the Île d'Oléron, we discarded commercial devices in favor of a 12 V battery with which we could easily supply 1 A.

5 This is a classic formula that can be found everywhere. However, it is incorrect in terms of probability theory. The "right formula", for independent measurements, is: $\sigma_{\frac{\delta\rho_a}{\rho_a}} = \sqrt{\sigma_{\frac{\delta\Delta V}{\Delta V}}^2 + \sigma_{\frac{\delta I}{I}}^2}$. Here, σ_α designates the standard deviation(#) for the quantity α.

When there are few parameters, the mistake made in relation to the error is not catastrophic.

we need $|\Delta V|$ to be as large as possible, i.e. the greatest possible current, as these two terms are linked by Ohm's law to the total ground resistance. Therefore, unless we can increase the power, we should optimize this ground connection: we should sink the electrodes well and possibly pour some saltwater on them. However, this line of thinking stumbles over the very technological limitations of the device.

5.3.2. *Moving from direct to alternating current*

5.3.2.1. *Measurements with direct current*

The simplest type of equipment includes, besides coils and cabling, a direct current generator, linked to an ammeter and a voltmeter, both for direct current. The simplest kinds of direct current generators are still batteries.

A simple alkaline 9 V battery (thus, a 6LR61 type of battery) offers a capacity of slightly more than 500 mAh. With five batteries, for example, we can rely on a 45 V generator. If the current goes up to 100 mA, we can supply power for 5 h. In practice, this is a lot, as the transmission period for a measurement does not last more than 5 s. At this rate, we can take 3,600 measurements.

The protocol assumes that we will first of all measure SP without supplying any power, let us say V_{PS}, measuring then the differential tension between P_1 and P_2, let us say V_1, as well as the current I, as power is being supplied. Therefore, we will calculate:

$$\rho_a = K \frac{(V_I - V_{PS})}{I}$$

As SP may vary slightly (recent installation of the electrodes, etc.), it is wise to measure the SP before supplying any power, and to take another measurement afterwards, and finally to calculate the average. If the SP is negative, as it should be half the time (!), we take away a negative value. This is the same as adding a number: $-(-25) = +25$, as we will all agree.

If we can only rely on a unique multimeter, we take two measurements (besides measuring the SP): one with a voltmeter from P_1P_2 and the other in a series with the generator, connected to an ammeter.

Nothing can be easier than carrying out this type of electrical sounding (or other types of measurements), but the interpretation is another story, as we pointed out in Chapter 3.

5.3.2.2. Moving on to alternating current

Using alternating current allows us to get rid of spontaneous polarization, which involves direct current. Thus, we transmit a sinusoidal or square-wave signal, and what we measure is an alternating tension, preferably with an "rms" voltmeter. However, this has certain geophysical consequences as follows:

– We need to use a frequency low enough to avoid the "skin effect"(#). Let us calculate the penetration depth of an electromagnetic (and therefore electrical) field for a typical resistivity of $\rho = 100\ \Omega\,m$ and a frequency $f = 1$ kHz. The depth is equal to:

$$d = \sqrt{\frac{\rho}{\pi f \mu}}$$

We can consider a magnetic permeability μ close to that found in a vacuum: $\mu_0 = 4\pi \cdot 10^{-7}$ [Henry / m]. Thus, $d = 159$ m.

Consequently, if we "stretch" a sounding to more than $C_1C_2 = 100$ m, the skin effect occurs, leading to a minimization of the tension measured. This depth is equal to 460 m at 120 Hz, and we would need a sounding of more than 1 km to find actual disturbance (always for $\rho = 100\ \Omega\ m$).

Figure 5.4 generalizes this result.

– At all frequencies, we can find a component of induced polarization. It is generally less than a few percentage points (unless we have reasons to expect a high value, for example if we encounter metal pollution) and in most cases disregarded.

– There may be some traps linked to the fact that most often we lose the sign of the signal. We can show cases where the measured signal changes sign, logically leading to apparent resistivity values that must naturally be negative (see Figure 2.10). Strictly speaking, we should measure the component (in phase with the transmission) of the signal measured, which involves precisely keeping in mind the phase of the signal transmitted. This requires a reference point related to the current transmitted for the receiver. In practice, this is not always done. That being said, finding negative apparent resistivity values in the field generally means that we have inadvertently inverted P_1 and P_2. An alternating current device will not experience this.

– With a device designed to measure induced polarization, the measurement takes the sign into consideration, so that this issue does not arise.

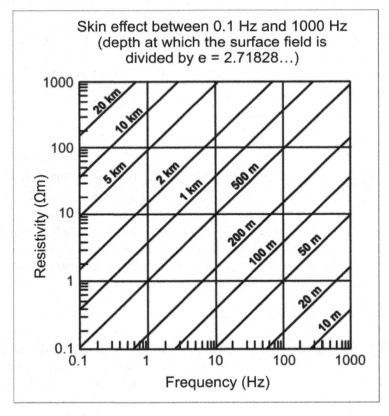

Figure 5.4. *The image shows the depth of penetration in relation to the frequency and resistivity of a homogeneous type of ground. This notion is merely theoretical, as the subsurface is not homogeneous*

5.3.3. *Alternating current transmitters*

5.3.3.1. *Popular existing systems*

In the literature, we can quote two articles published in the English journal "Everyday Practical Electronics". The first, written by Robert Beck in February 1997, can now be freely accessed, for example here:

http://www.geotech1.com/pages/geo/projects/erm1/erm1a_300.pdf

http://www.geotech1.com/pages/geo/projects/erm1/erm1b_300.pdf

The second article was written by John Becker and published in April 2003. It cannot be freely accessed yet, and it is called "Earth Resistivity Logger". While we wait for it to come into the public domain, we can find a few excerpts on this website:

https://fr.slideshare.net/jplateado/earth-resistivity-logger-john-becker

These systems are undoubtedly effective and also implement a synchronous detection that allows us to keep the sign when the apparent resistivity is negative (this should not happen often!). A synchronous detection can also limit "out-of-band" noise.

However, these outlines are all designed for low tensions, so that even a system based on a few 9 V batteries and working on direct current will allow us to go deeper.

Here, we put forward two outlines of regulated, alternating current transmitters and we discuss their advantages and possible alterations. We encourage readers to go through the safety measures carefully before starting to build these devices.

5.3.3.2. *The regulated current transmitter "MINELEC1"*

Its outline is illustrated in Figure 5.5. A classic multivibrator supplies a transformer 220 V →2X7 V, 10 VA that has been inverted and turned into a voltage step-up device. It is supplied by a series of cell batteries (LR6) reaching 12 V (eight elements for alkaline ones). The tension generated is then rectified with a diode bridge and filtered. After being filtered, this tension supplies the electrode C_1. The limiting factor for both tension and power derives from the use of the only two low-power transistors 2N2905, which we will equip with heat sinks. In practice, we can expect to reach around 100 V. We could also rely on a series of around 10 9 V batteries.

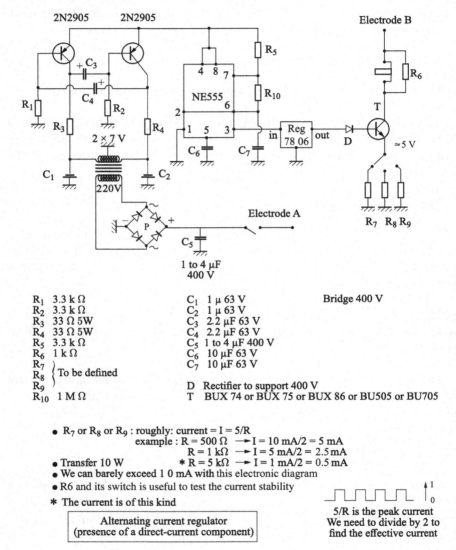

R_1	3.3 k Ω	C_1	1 μ 63 V
R_2	3.3 k Ω	C_2	1 μ 63 V
R_3	33 Ω 5W	C_3	2.2 μF 63 V
R_4	33 Ω 5W	C_4	2.2 μF 63 V
R_5	3.3 k Ω	C_5	1 to 4 μF 400 V
R_6	1 k Ω	C_6	10 μF 63 V
R_7		C_7	10 μF 63 V

Bridge 400 V

R_7
R_8 } To be defined
R_9
R_{10} 1 M Ω

D   Rectifier to support 400 V
T   BUX 74 or BUX 75 or BUX 86 or BU505 or BU705

- R_7 or R_8 or R_9 : roughly: current = I = 5/R
  example : R = 500 Ω →I = 10 mA/2 = 5 mA
            R = 1 kΩ →I = 5 mA/2 = 2.5 mA
- Transfer 10 W   * R = 5 kΩ → I = 1 mA/2 = 0.5 mA
- We can barely exceed 1 0 mA with this electronic diagram
- R6 and its switch is useful to test the current stability
* The current is of this kind

5/R is the peak current
We need to divide by 2 to
find the effective current

| Alternating current regulator |
| (presence of a direct-current component) |

Figure 5.5. *Outline of the regulated current transmitter MINELEC1*

Besides, a NE555 produces a signal whose frequency is:

$$f = \frac{1.44}{(R_5 + 2R_{10})C_7}$$

for a cycle ratio (short square-wave in relation to the long square-wave) equal to:

$$D = \frac{R_{10}}{R_5 + 2R_{10}}$$

For the values indicated in the circuit, the frequency is in the range of 72 Hz and the cycle is very close to 1, which corresponds to a virtually square signal.

The third pin of the 555 is plugged into a regulator 7806 that only supplies its 6 V for a positive square-wave of the 555. This signal reaches the high tension transistor by means of a safety diode (400 V, maybe useless, but when the transmitter of the transistor is up, the HT of the generator – around 200 V – risks damaging the 555). The transistor BUX84, or any equivalent transistor, is turned into a regulator (it must also support 400 V; these are transistors that were once used to sweep cathode ray tube televisions). The tension at the emitter is close to $6 - 2 \times (0.5) = 5\,\mathrm{V}$. The resistance of the emitter, here R_7, R_8 or R_9, determines the emitter current, very close to that of the collector, (roughly β, gain of the transistor), so that we obtain:

$$I \cong \frac{5}{R_{7(or\,8\,or\,9)}}$$

This is the current for a positive alternation of the 555. When in standby, no current is transmitted. As the cycle is a square-wave cycle close to 1, the effective current is half this value, namely

$$I_{eff} \cong \frac{2.5}{R_{7(or\,8\,or\,9)}}$$

In most cases, this transmitter can be used according to the following table:

For a maximum output tension of around 100 V	$R = 500\,\Omega$	$R = 1\,k\Omega$	$R = 5\,k\Omega$
Effective current	5 mA	2.5 mA	0.5 mA
Peak current (positive alternation)	10 mA	5 mA	1 mA
Power	10 W	5 W	1 W
Maximum ground resistance (sum of the two electrodes)	$R = 10\,k\Omega$	$R = 20\,k\Omega$	$R = 100\,k\Omega$

The observations are as follows:

– Regulating an alternating current requires us to manage two timespans: that of the current itself and that of the regulation. The solution we have chosen here involves chopping an already regulated current.

– The device is adapted for high ground resistance. It is designed to be used with arrays that employ a stand, as is the case in Figure 5.6, since it often happens that the "rapid" contact made with this method is not perfect. It is wise to test the system by only positioning or sinking the electrodes very slightly to assess the quality of the contact. Weak currents are also adapted for such a process. With a pole–pole array, where a = C_1P_1, and in a type of ground at $100\ \Omega m$, with an effective current of 0.5 Ma, the potential is

$$V = \frac{\rho I}{2\pi a} = \frac{100 \times 0.5}{6.28} = 8\ mV .$$ This is a lower limit (in relation to the noise, as of its own it can be amplified!). Let us consider this example to assess the ground resistance and review what we can do. For a post of radius r = 5 mm, sunk l = 5 cm into the ground (discarding that it has a point that makes it easier to sink it), calculating the ground resistance yields (see the section

2.2.2 about ground resistance): $R = \dfrac{\rho}{2\pi L}\left[\log_e\left(\dfrac{4L}{r}\right) - 1 \right] = 663\,\Omega .$ To find

the total resistance, we need to double this value. We are quite far from the maximum ground resistance value for an effective current of 5 mA, which makes the potential go up to 80 mV. If by chance the ground resistivity reaches $1,000\ \Omega m$ and resistance is accordingly $2 \times 6,630 = 13,260\ \Omega$, we exceed the tension capacity of the device. It is by using common sense or monitoring the tension of the generator that geophysicists will then employ another range.

– The current as it is transmitted involves a direct current component. This does not matter. When received, the peak–peak amplitude of the signal will in most cases be smaller than the SP. Our goal is to measure the alternating part of the tension rather than its direct component (otherwise, moving on to alternating current would be useless).

– We can add a current measurement to the transmitter, for example with a shunt. However, the tension at the shunt will be much higher than the tensions at P_1 and P_2. This implies that it is *indispensable* to completely disconnect the measurement of the current from the measurement of the tension. Unless we use two galvanically independent receivers (independent supplies with independent masses), the only solution is to rely on an isolation amplifier[6]. If the output stage of this amplifier can be supplied by the receiver, its input stage will require in turn an independent (floating) supply from the transmitter. Consequently, we need to block any possible galvanic connection between the high tension circuit (in the end, everything is relative) and the tension measuring circuit.

Figure 5.6. *Implementing MINELEC1 on the Aulus site (the author is on the right). We can see the stand, its electrodes and cables, an acquisition case and measuring tape being used on the ground*

6 For example, an ISO124P built by Texas Instrument. We should point out that another circuit like AD202 can supply the input stage on its own, internally and ensuring insulation. Thus, this system is an elegant (if somewhat more expensive) solution for measuring current.

Measuring the current allows us to stop regulating the current. Instead of being defined by its design, the current is measured. This is undoubtedly an advantage. We only need to set the emitter resistance of BUX84 at 0 and, just to be safe, attach BUX84 to a heat sink. Finally, by adopting an inverter circuit such as that of MINELEC3, described further down, and with this system that cuts the power, we can obtain higher tension and power.

5.3.3.3. *The receiver of MINELEC1*

At the transmitter frequency, i.e. around 70 Hz, a classic multimeter in alternating voltmeter mode will yield excellent results. However, these results can only be read graphically and must be reported on a page of our field notebook. A second multimeter, used as an alternating ammeter, will allow us to read the current. Then, we will use an Excel file or another type of file to enter the coordinates of the point, the tension and the current to calculate an apparent resistivity. We will then bring all this back to the office.

A receiver more adapted for digital acquisition requires a small electronic system. Figure 5.7 puts forward an outline.

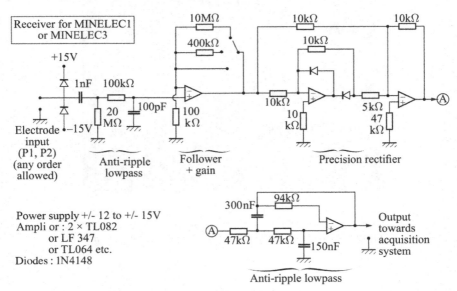

Figure 5.7. *Receiver for MINELEC1 and MINELEC3*

We did not set up the input as an instrumentation amplifier on purpose. The electrode P_2 is directly grounded. The input impedance is greater than $20\,M\Omega$. As the symmetrical supply of the system from (–9 V, 0 V, +9 V) to (–15 V, 0 V, +15 V) is isolated from ground, this does not pose any problem and consequently the risk of encountering common-mode problems is eliminated. The electrode P1 is plugged into the input amplifier (1/4 of LF347 or TL064, for example – it is not that significant) by means of two safety diodes and a passive passband filter, which eliminates the direct current component and the very low frequency (VLF) signals, and allows the op amplifier input (+) to discharge potential static. The same first op-amp can regulate a gain, which is useful for small signals. The second and the third op-amps are mounted with two diodes, preferably matched, set up as a precision rectifier. The last op-amp of the series is set up as a low-pass filter and eliminates the output ripples of the precision rectifier. According to the tension expected for the following analog-to-digital converter, we can certainly add a divider bridge or any other kind of adjustment, and, if we wish, we can add a led, a galvanometer, or a buzzer to indicate saturation.

We can simply calibrate with a multimeter. With an input signal (coming from the ground or a resistor bank linked to the transmitter, or with a signal generator), we only need to set up the multimeter as an alternating voltmeter in line with the entry, and then set the multimeter in direct current mode for the output. We can obtain the gain as:

$$\text{Gain} = \frac{V_{output}^{=}}{V_{input}^{\sim}}$$

With the acquisition program, dividing the raw measurement by the gain allows us to find out the effective input voltage, which is used to calculate the apparent resistivity.

We can use the same outline to measure the current. To this end, the transmission line goes through a shunt box that includes the isolation amplifier whose input must be supplied independently of the rest of the electronic system. It is recommended to use a resistance of 10 or 50 Ω for this shunt since, given that the currents are most often weak, we will obtain an equal gain before the isolation amplifier. The acquisition program will take this gain into account by dividing the value measured by this resistance. Thus, together with the receiver gain, we need to consider two gains.   •

5.3.4. *"Hight voltage" transmitters*

By "high voltage" we mean dangerous voltages in the field, in this case ranging from 50 to 300 V. Any higher voltage should definitely be avoided by amateurs; the risks are too real and special, and high-quality insulated material, which is neither common nor easy to find, is required.

We remind readers that the operator cannot be actively protected when carrying out electrical prospecting and we *ask them to read and reread the chapter dedicated to the safety measures*. The currents stated in this section as well as the following may exceed 100 mA, which can cause death in a few seconds when they flow through our body.

5.3.4.1. *With an inverter found on the market*

We can find very easily some "inverters" or "converters" that transform the 12 V of a car batter into a "household" 230 V.[7] Let us add a switch, a pushbutton, a multimeter set up as an alternating ammeter (with a range of 200 mA) and another set up as an alternating voltmeter (with a range of 200 mA). With this equipment, the coils and the electrodes, we have what we need to carry out an electrical sounding involving a C_1C_2 that may be several hundred meters long.

The transmitter, its portable 12 V battery ("NP7" type) and the pushbutton (or better, two pushbuttons!) can be placed in a "transmitter" box. The pushbuttons are used as a safety measure: no current is flowing when we don't push both buttons at the same time.

Some individuals at the "Practica Foundation" have understood this trick, and I congratulate them:

https://www.practica.org/wp-content/uploads/2016/05/Manual-Bedrock-Kit-EN-MWBP0221216.pdf

(see their Website: https://www.practica.org/)

---

7 In Europe, the rated voltage is 230 V (effective) instead of "220", which is no longer used. The map we can see here http://northbysouthwest.fr/wp-content/uploads/2011/10/ voltage_pays.png shows which voltages are used all over the world.

5.3.4.2. A regulated current transmitter: MINELEC3

This is a regulated alternating current transmitter (Figure 5.8).

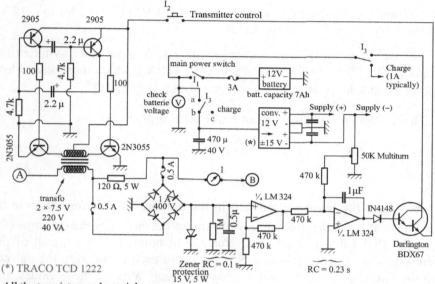

Figure 5.8. *A regulated alternating current transmitter (100 mA maximum)*

The alternating current is produced by an inverter ballasted with two power transistors (mounted on heat sinks) 2N3055. The output coil of the inverted transformer (input: 2×7.5 V, output 230 V) directly supplies the electrodes C_1 and C_2 by means of a shunt S (typically $S = 100 \,\Omega$). The regulation involves a proportional-integral type of feedback control. Thus, the shunt tension is corrected by a diode bridge that accepts 400 V, followed by a passive low pass with a time constant RC = 0.1 s. It is connected into a follower (of gain 2). We add a negative reference taken from a multiturn potentiometer to the output tension of this op-amp. We insert a safety diode behind this integrator before connected to in a power Darlington BDX67, which must be properly cooled. The feedback control, as we can see, is connected to into the supply voltage of the inverter.

All this is supplied by a 12 V battery, but we must also create a supply that is perfectly insulated from the high voltage to supply the op-amps of the feedback control system. There are two solutions to achieve this:

– either we can rely on a set of batteries (–12 V, 0 V, +12 V) to (–15 V, 0 V, +15 V) independent of the battery;

– or we can use a DC-DC converter with a 12 V input and a symmetrical output (–12 V, 0 V, +12 V) to (–15 V, 0 V, +15 V), which must have a low coupling capacitance and good insulation (1,000 V or more).[8]

The maximum effective current transmitted is given by:

$$I_{eff}^{max} = \frac{2}{\pi} \frac{R_0}{R_1} |U_{ref}| \frac{1}{S}$$

where U_{ref} is the reference tension. In the outline, with the multiturn at its maximum, it is the *negative* supply tension of the circuit of the op-amps (for example 15 V) that is used as a reference to build the error signal of the feedback control. The resistance S is the shunt used to measure the current and the two resistances R_0 and R_1 may be equal. With all these values, the max current is 96 mA. The tension at the terminals of the transmitter often goes up to 300 V (this threshold should not be crossed according to the isolation performance of the transformer, otherwise a transformer 2 × 9 V ⇔ 230 V, or a transformer designed for a higher voltage, is more appropriate).

Important precaution: When the circuit is open, the feedback control will increase the output tension to its maximum. If we happened to extract an electrode with our hands...sudden death would be a possibility! It is convenient to place the pushbutton (or two pushbuttons) on the supply line that goes from the transmitter of the Darlington to the inverter, as pointed out in the outline. No oscillation, no current, no risk involved. We can also rely on the on-off switch as a safety measure. Thus, when carrying out an experiment, we turn on the power when everything is ready and the operators are far from the electrodes, and then we press the pushbutton. In a very short time, the tension reaches its maximum and then completely

---

8 Brands such as TRACO or MURATA have specialized in this type of converter. If readers wish to look them up in Google, the keyword is, for example, "DC/DC CONVERTER DUAL OUTPUT".

stabilizes, dropping in ½ a second. Then, we play with the multiturn to obtain the current desired, verified by an ammeter.

We can use the aforementioned receiver or a simple voltmeter set in alternating current mode for the reception of this inverter. As for measuring the current, what we said about MINELEC1 applies here as well.

5.4. Equipment for induced polarization

In this volume, we only put forward the simplest system, which measures two frequencies. We simply have to supplement the measurement at a frequency f by a measurement at another frequency F (for example 5 Hz and 100 Hz, so we need a frequency commutator at the transmitter and possibly at the receiver). To take into consideration the filters of the receivers and make the process easier, the best way is to calibrate each of the frequencies on a non-polarizable circuit or, in other words, on a bank of pure resistances. With this calibration, the response will be the same for the two frequencies, changing instead in a polarizable medium.

An Acquisition System Designed for the Electrical Prospection of Soil

6.1. The presentation of the "open source" project

The implementation project presented to the reader in this chapter is designed for the electrical prospection of the subsoil through resistivity measurements. This device is "opensource", so that readers will be free to modify the program, which includes adapting it according to their own needs or adding new features.

Our goal is to point out to the reader all the technical resources required to implement this tool, detailing each step of its design process as well as the numerous related technologies.

Electrical prospecting involves transmitting currents to a pair of electrodes planted in the ground and measuring potentials in other points of the surface. This tool is designed only for the acquisition of analog values, such as intensity and tension. Afterwards, the signals are converted into a digital format and saved and stored so that they can be analyzed with a specialized geophysical program. The issue is to determine the spatial variation of resistivity in the subsoil.

The sources of the software can be found on the GitHub platform at https://github.com/MuhlachF/Magneto.

6.2. The preliminary study

6.2.1. *The analysis*

The building block model allows us to identify five components as follows:

– a chip that carries out the analog–digital conversions (analog–digital converter [ADC]);

– a microcontroller that processes information;

– a backup unit;

– a man–machine interface, which allows us to interact with the system (screen, keyboard, etc.);

– modules whose task is to timestamp and geolocate the data, handy for the interpretation of the results.

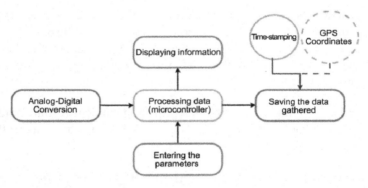

Figure 6.1. *Functional outline of the project. For a color version of this figure, see www.iste.co.uk/florsch/geophysics1.zip*

6.3. Choosing the components

6.3.1. *The microcontroller*

The microcontroller is an integrated circuit whose main task is:

– to process the information;

– to drive the peripheral devices connected to its pins.

It includes six components as follows:

– the central processing unit, which manages the digital streams and sends orders in relation to the program run. It also manages the inputs–outputs, addresses storage areas (reading and storing information) and responds to interruptions (interrupt request);

– the system bus, which connects the processor with internal hardware devices;

– a static read access memory (SRAM) storage area, where all the temporary data required for the calculations are stored in a volatile area;

– a FLASH working memory, whose task is to conserve the program even if the card is no longer supplied;

– an EEPROM (Electrically Erasable Programmable Read-Only Memory) memory, which can be erased and programmed electrically (around 100,000 write cycles). Durable data (such as the configuration elements) is saved here;

– input–output ports will allow the microcontroller to dialog with the external world, especially with all the peripheral devices that will be connected to it.

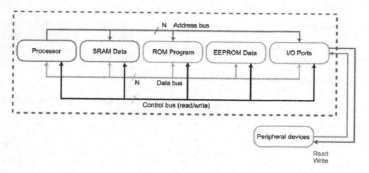

Figure 6.2. *Functional outline of a microcontroller. For a color version of this figure, see www.iste.co.uk/florsch/geophysics1.zip*

Arduino boards, which rely on Atmega architecture, can be easily found at an affordable price. We have chosen the Mega2560 version, which includes enough memory to run the program. Its smaller counterpart, the UNO model, can still be used, provided that we accept not to acquire GPS data.

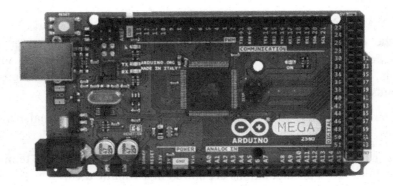

Figure 6.3. *The Mega2560 version of the Arduino board*

Its technical features are given in the table below:

Characteristics	Value
Microcontroller	ATmega2560
Clock frequency	16 MHz
Operating voltage	5 V
Range of recommended input voltages (limit values)	7–12 V (6–20 V)
Number of digital ports	Fifty-four of which 14 can work in PWM (pulse width modulation)
Number of analog ports	Sixteen analog inputs
Maximum current transmitted per pin	40 mA
Memory	256 Ko Flash, 8 Ko SRAM, 4 Ko EEPROM
Approximate price	€45

6.3.2. *The communication protocols of the microcontroller*

Most extension cards used for this purpose employ inter-integrated circuit (I2C) and serial peripheral interface (SPI) protocols. These protocols can reduce the number of electrical wires, as the data stream is transmitted with the same bus.

These SPI links work according to the master–slave principle and use four logical signals:

– SCLK (Serial Clock) → clock generated by the master module;

– MOSI (Master Output, Slave Input) → driving the slave modules;

– MISO (Master Input, Slave Output) → collecting the information deriving from the controlled modules;

– SS (Slave Select) → selecting the component with which the module wants to communicate. The number of SS lines is the same as that of slave modules.

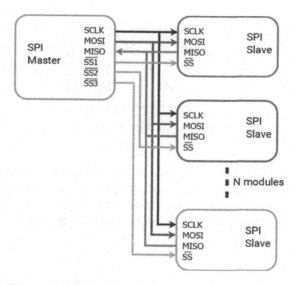

Figure 6.4. *The SPI protocol. For a color version of this figure, see www.iste.co.uk/florsch/geophysics1.zip*

Put forward by Philips, the I^2C protocol can address 128 components with only two lines:

– the bidirectional Serial Data Line (SDA) allows the components to exchange information;

– the Serial Clock Line (SCL), which is also bidirectional, carries the clock signal.

NOTE.– With the I^2C protocol, each module is assigned an address.

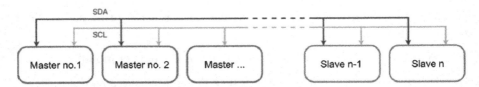

Figure 6.5. *The I²C protocol*

6.3.3. *The analog–digital converter*

An ADC is an electronic device that transforms analog signals into digital data.

Figure 6.6. *The ADC symbol*

The precision of an ADC depends on its resolution (or bit count). For a material with a 10bit conversion (Arduino), the analog quantities are coded following 2^{10} states, i.e. values from 0 to 1023. With a reference of 5 V for Vref+ and 0 V for Vref–, the quantum* is equal to 4.88 mV.

$$q = \frac{V_{ref+} - V_{ref-}}{2^n}$$

* The quantum corresponds to the smallest input voltage change that will lead to an increment or decrement of the output conversion digital value.

The perfect components do not exist, and several parameters affect the quality of the conversion. In practice, chips are characterized by more or less significant discrepancy errors (offset error, gain error, INL, DNL), which result from several factors such as the amplification of the signals, temperature, etc. These errors should be taken into account when choosing a component: the closer it is to its ideal model, the higher its price will probably be.

The ADC inputs can generally be used in two ways:

– as single-ended inputs, the tension is measured with only one wire, using the mass of the assembly as a reference point. Disturbances are recorded together with the signal;

– in differential mode, the tension measured is that among two wires; the interferences, present on both paths, are thus disregarded.

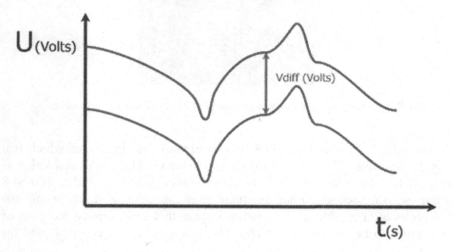

Figure 6.7. *The measurement in differential mode*

The differential mode is recommended for long lines or lines that are likely to present interferences. This is why we will use it to take our measurements.

The analog inputs of the Arduino model may work in this mode. Despite this, their resolution and absolute accuracy are relatively limited. Therefore, we need to use specialized chips.

Offered in a surface-mounted device format, this material is first of all designed for professional technology integrators that pay attention to the place occupied by a circuit in a given volume. Several companies have made the most of the craze started by the makers to provide to their customers these components in a "Breakout Board" format, which is more adapted for the connector boards used in our experiments.

Thus, a company called Adafruit offers the ADS1115 converter (16 bits of resolution) in its "Breakout Board" version:

https://www.adafruit.com/product/1085

Figure 6.8. *The Breakout Board version of the ADS1115 converter*

The ADS1115 chip is an ADC that integrates four inputs, of which two may be multiplexed to work in differential mode. The 16-bit resolution is adapted for electrical sounding. The maximum operative speed is 860 SPS (samples per second). Four modules may be addressed by using the I^2C protocol. The chip also includes a stage that can amplify the gain of the signals to be measured (FPGA). This converter is perfectly adapted for our project.

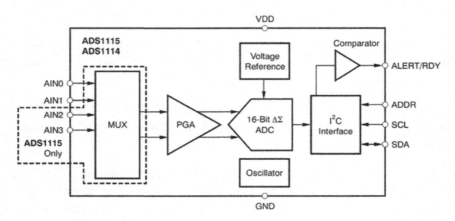

Figure 6.9. *Internal circuit of the ADS1115 (source: Adafruit)*

6.3.4. *From the man–machine interface to data saving*

The interface is based on:

– a peripheral device produced by Sparkfun (VKey PRT12080), whose technology is modeled on the principle of voltage dividers. Each button corresponds to a different analog level. Other types of keyboards (I^2C) may be used, provided that we modify the program;

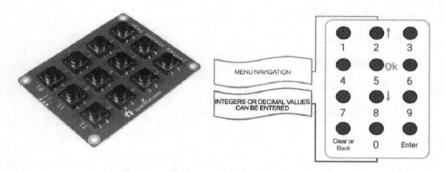

Figure 6.10. *Features of the keyboard*

– an LCD screen displaying two 16-character lines, which displays the parameter information as well as the menus. It may be possible to activate a backlight according to the users' preferences or the weather conditions.

Figure 6.11. *An I^2C LCD screen in the Grove format produced by Seeed*

6.3.5. *The RTC module and data saving*

The real-time clock (RTC) is a card in the Grove format built by a company called Seeed. Its main component is a DS1307 circuit. It displays the hour to the nanosecond. Seven registers can be read:

– Seconds (from 0 to 59);

– Minutes (from 0 to 59);

– Hours (from 0 to 23);

– Days of the week (from 1 to 7);

– Dates (from 1 to 31);

– Months (from 1 to 12);

– Years (0 corresponds to the year 2000).

Figure 6.12. *The I^2C RTC clock (Grove format) built by Seeed*

A microSD card reader, in the SPI format, is used to save all the data as it is acquired.

Figure 6.13. *SD card reader (SPI format)*

6.4. Outline of the layout of components

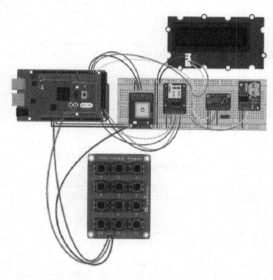

Figure 6.14. *The outline created with the program Fritzing. For a color version of this figure, see www.iste.co.uk/florsch/geophysics1.zip*

6.4.1. *Connections*

Peripheral device	Connection and power supply	Protocol used
Sparkfun 12-key keyboard	+ → 5 V and GND (mass) Out → Arduino A0 Input	Analog link
Option (GPS Adafruit)	Vin (5 V) and GND (mass) TX → RX1 Arduino RX → TX1 Arduino	Serial connection
microSD card reader (Adafruit)	+ -> 5 V and GND (mass) CLK → port 52 (SCK) DO → port 50 (MISO) DI → port 51 (MOSI) CS → port 53 (SS)	SPI
ADS1115 (Adafruit)	VDD → 5 V and GND (mass) SCL → port 21 (SCL) SDA → port 20 (SDA) ADR -> GND	I²C Address: 0 × 48 (1 001 000)
RTC Module (Grove)	VDD → 5 V and GND (mass) SCL → port 21 (SCL) SDA → port 20 (SDA)	I²C

6.4.2. *List*

Name	Link	Indicative price (€)
Arduino Mega2560 Board	https://www.arduino.cc/en/Main/ArduinoBoardMega2560	40
Analog keyboard Sparkfun	https://www.sparkfun.com/products/12080	15
GPS 66 ch Adafruit + Serial SMA SMA960	https://www.adafruit.com/product/746	55
MicroSD card reader Adafruit	https://www.adafruit.com/product/254	8
16-bit Analog-Digital Converter Adafruit	https://www.adafruit.com/product/1085	15
RTC Module Seeed	http://wiki.seeed.cc/Grove-RTC/	8
LCD I²C Screen Seeed	http://wiki.seeed.cc/Grove-LCD_RGB_Backlight/	14

6.5. Preparing the microSD card

The Arduino reader accepts cards in a microSD format regardless of their capacity. The library developed for the Arduino reader requires a maximum addressable size of around 2 GB. Now, it has become difficult to come across cards of this type that are smaller than 2 GB.

To overcome this obstacle, a partition management tool like AOMEI Partition Assistant must be used. This program can be freely used for private and non-commercial use:

http://www.aomeitech.com/aomei-partition-assistant.html

Once we have run the program, it lists all the storage peripheral devices detected by the system. Then, let us click on the area of the disk that corresponds to our SD card and choose, in the menu displayed, the tab "Create partition".

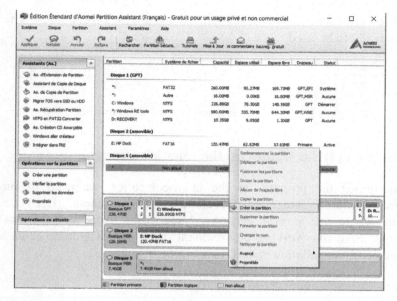

Figure 6.15. *The free version of the program Partition Assistant (AOMEI)*

A 1 GB partition represents free space to save our data. Let us choose FAT as the file system.

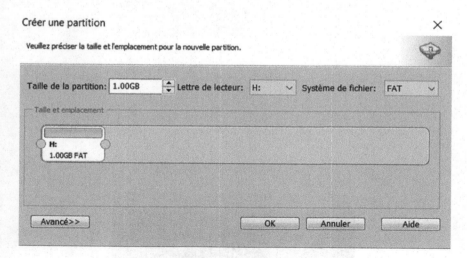

Figure 6.16. *A 1 GB partition in FAT*

Let us click on "Apply" to take the changes into consideration. The card is then ready to be used.

Figure 6.17. *The SD card now has a 1 GB partition in FAT*

6.6. Running the program on the microcontroller

To upload the program to the microcontroller, we need to run the program Arduino Integrated Development Environment (IDE) on our system. There are several versions. Necessarily, there will be one that corresponds to our working environment: https://www.arduino.cc/en/Main/Software. The installation instructions can be read here:

https://www.arduino.cc/en/Guide/HomePage

Once we have installed the program, we can run the editor. The window is shown in Figure 6.18.

Figure 6.18. *The Arduino IDE*

Two idle functions (namely, without code lines) appear in the IDE:

– **setup** () includes the instructions followed as we open the card or when we reboot it after pushing the RESET button. The pin configuration (INPUT, OUTPUT) and that of the connected modules will be displayed here;

– the **loop** () function includes the program which is run on loop.

The program developed relies on several libraries, which must be installed before compiling the program and uploading it to the microcontroller:

– For the GPS: https://github.com/adafruit/Adafruit_GPS.

– For the converter: https://github.com/adafruit/Adafruit_ADS1X15.

– For the LCD screen: https://github.com/Seeed-Studio/Grove_LCD_RGB_Backlight.

– For the RTC module: http://wiki.seeed.cc/Grove-RTC/.

NOTE.– The SD card library is already part of the IDE environment.

Example: To download a library from GitHub, click on "Clone or download" and then on "Download ZIP".

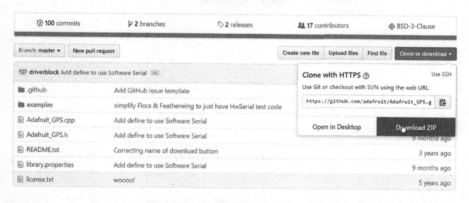

Figure 6.19. *The GitHub service*

We now have a compressed version of the library: Adafruit_GPS-master.zip.

To open the libraries with the Arduino editor, here is what we need to do: in the "Sketch" menu of the IDE, choose "Import library" and then "Add .ZIP Library". Now we only need to choose the zip file that corresponds to the library we have chosen.

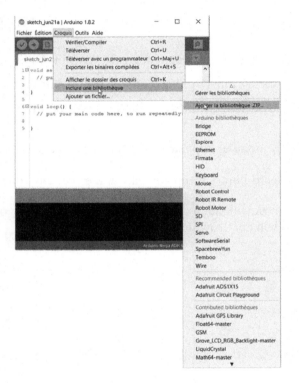

Figure 6.20. *Installing a library*

Now, we only need to download the sources of the program from the GitHub platform. This service allows us to share open source projects and create a community that can in turn suggest changes or improvements. This operation is called "pull requests": https://github.com/MuhlachF/Magneto.

After extracting the file on our hard drive, let us open the program with the Arduino IDE.

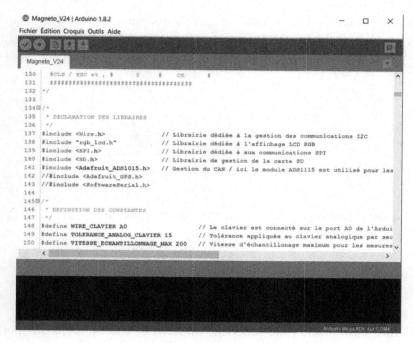

Figure 6.21. *Version 0.24 of the program "Magneto"*

6.6.1. *Additional configuration settings*

The device is initially designed to work with tensions between −2.048 and 2.048 V. However, we can modify this parameter to widen or narrow this range. We should be aware that any change affects the characteristics of the circuit: a high gain decreases the range measured, but it increases the precision of the conversions. The following values given in the table may be used.

PGA setting	FS (V)
2/3	±6.144
1	±4.096
2	±2.048
4	±1.024
8	±0.512
16	±0.256

We will have to modify two lines of the program: the first involves the gain, which is applied to the PGA stage of the converter.

```
/*
 * CONFIGURATION CAN ADS1115
 */
// ads.setGain(GAIN_TWOTHIRDS);  // 2/3x gain +/- 6.144V  1 bit = 3mV     0.1875mV (default)
//ads.setGain(GAIN_ONE);         // 1x gain   +/- 4.096V  1 bit = 2mV     0.125mV
ads.setGain(GAIN_TWO);           // 2x gain   +/- 2.048V  1 bit = 1mV     0.0625mV
// ads.setGain(GAIN_FOUR);       // 4x gain   +/- 1.024V  1 bit = 0.5mV   0.03125mV
// ads.setGain(GAIN_EIGHT);      // 8x gain   +/- 0.512V  1 bit = 0.25mV  0.015625mV
// ads.setGain(GAIN_SIXTEEN);    // 16x gain  +/- 0.256V  1 bit = 0.125mV 0.0078125mV
ads.begin();
```

Figure 6.22. *Code lines related to the adjustment of the CAN gain*

The second involves the variable that contains the multiplier:

```
/*
 * VARIABLES / CAN
 */
Adafruit_ADS1115 ads;            // Déclaration du convertisseur
//double multiplier = 0.125F; /* ADS1115 @ // 1x gain   +/- 4.096V (16-bit results) */
double multiplier = 0.0625F;  /* ADS1115 @ // 2x gain   +/- 2.048V (16-bit results) */
```

Before downloading the program, we need to choose the COM port to which the Arduino module is connected in the "Tools/PortComx" menu and the type of card (here, Arduino Mega ADK) in the "Tools/Type of card" menu.

The code lines may now be compiled and sent to our microcontroller by clicking on the arrow displayed in the editor shortcuts:

6.7. The program and its menus

6.7.1. *Initialization steps for the device*

As soon as the device is on or rebooted, the configuration settings are loaded into memory, the modules are opened and the integrity of the storage unit is verified.

Once these control operations have been carried out, the program displays the main menu.

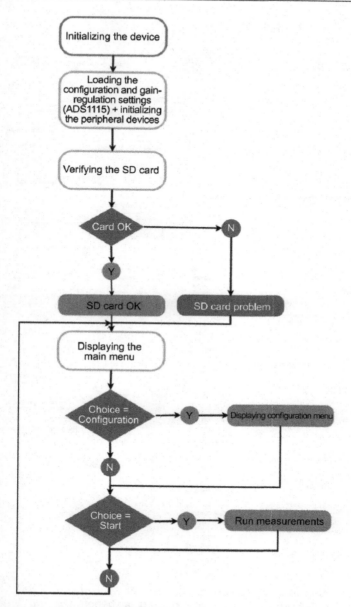

Figure 6.23. *A flow chart illustrating the main steps of the program: from the initialization to the menus displayed. For a color version of this figure, see www.iste.co.uk/florsch/geophysics1.zip*

6.7.2. *Presentation of the menus*

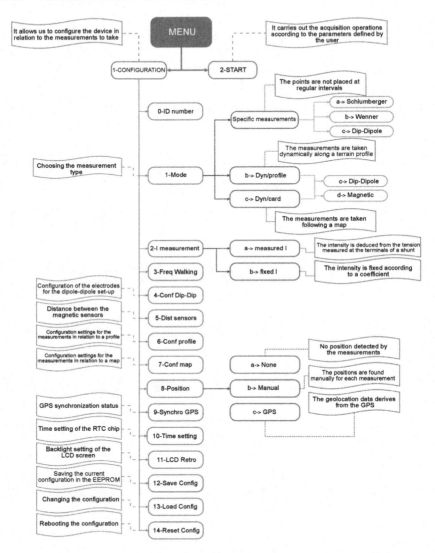

Figure 6.24. *The program menus. For a color version of this figure, see www.iste.co.uk/florsch/geophysics1.zip*

6.8. A practical example of how the device is used

6.8.1. *Preliminary configuration steps*

Here are the steps we should follow when carrying out the electrical prospection of a type of ground following the Schlumberger model.

First, we need to name our data:

In the menu "1 – Configuration", let us choose "0 – ID number", and then let us confirm (key 5).

The ID number can include up to six figures. Let us enter the value "1" and then confirm. As the document "1" cannot be found, it is created.

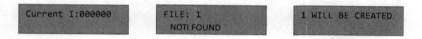

In the configuration menu, with the arrows (key 2 and 8), let us choose the option "1 – MODE" and then the Schlumberger model as a model.

Observation: An asterisk appears before the options that have already been chosen.

As for the current transmitted to the ground, we may find two scenarios:

– either it is measured with a shunt, so that it is necessary to specify a conversion factor;

– or the intensity generated is known beforehand and it is enough to enter its value in the program:

- in the "configuration" menu, let us choose "2-I measurement" and then "b→ fixed I", then let us enter the intensity in mA and confirm.

Finally, option "8" allows us to specify the type of position used.

6.8.2. *Taking the measurements*

Let us go back to the main menu (key 10) and choose "START" to take the measurements. In the meantime, a screen will display a summary of the options chosen. Let us simply confirm this.

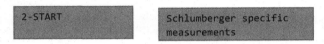

In Schlumberger mode, the distance P_1P_2 is fixed, unlike the distance C_1C_2 which varies according to the sounding.

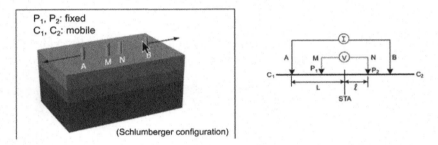

Figure 6.25. *Electrical prospection in Schlumberger mode*

To leave acquisition mode (to modify a parameter, complete the operations, etc.), let us enter "1" and then confirm.

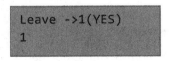

If we wish to record new data in the same field, we only need to go back to the main menu, choose the option "START", and then enter "1":

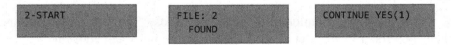

Figure 6.26. *The first tests are carried out in the field. In the foreground, we can see the measuring device model, the current generator and the battery. For a color version of this figure, see www.iste.co.uk/florsch/geophysics1.zip*

The first inputs involve the distance $0P_1$ and then $0C_1$, the values are in cm:

Distance(cm) OP$_1$:	Distance(cm) OC$_1$:

A potential input error may be corrected with the key CLEAR.

The measured values are displayed live. We should press "OK" to save these values on the SD card. The apparent resistivity is then shown, afterwards the value obtained is displayed:

U1: 1460.7500mV	Rhoa (approx): 4588967.0 Ohms.m

To leave acquisition mode (to modify a parameter, complete the operations, etc.), let us enter "1" and then confirm.

```
Leave ->1(YES)
1t
```

If we wish to record new data in the same field, we only need to go back to the main menu, choose the option "START", and then enter "1":

```
2-START
```

```
FILE: 2
   FOUND
```

```
CONTINUE YES(1)
```

6.8.3. *The result file*

Once we have inserted the card in a reader, the documents appear in the explorer window and a simple text editor shows the data recorded.

```
1   IDENTIFIANT DES MESURES : 2
2   MESURE SCHLUMBERGER
3
4   CONSTANTES DEFINIEES :
5   NOMBRE DE MESURES EFFECTUEES : 2
6   INTENSITE I2 FIXEE :32.00mA
7
8   ID MESURE, DISTANCE AB(cm), DISTANCE MN(cm), RHOA(Ohms.m), TENSION U1(mV), TENSION U2 (mV), COUTANT I2(mA)
9   1,6,3,170.943760,773.875000,0.000000,32.000000
10  2,98,3,60743.438000,773.812500,0.000000,32.000000
11
```

Figure 6.27. *The content of the file*

Figure 6.28. *Example of finalised version of the device.*

Appendix

Practical Links, Programs and Material

A.1. Modeling and interpreting vertical electrical soundings

A.1.1. *Commercial or semicommercial software (sometimes available in demo or trial version)*

EarthImager 1D
> https://www.agiusa.com/agi-earthimager-1d-ves

IPI2win
> ftp://www.geol.msu.ru/pub/geophys/ipi2win.htm

IX1Dv3
> http://www.interpex.com/ix1dv3/ix1dv3.htm

QWSELN
> http://www.sisyphe.jussieu.fr/~jtabbagh/pages/qwseln.htm

SensInv1D
> http://www.harbourdom.de/sensiv1d.htm

SPIA
> http://hgg.au.dk/software/spia-ves/

WINSEV
   http://www.wgeosoft.ch/PDF/WINSEV_data.html

ZONDIP1D
   http://zond-geo.ru/english/zond-software/ert-and-ves/zondip1d/

A.1.2. *Free, non-commercial software*

AarhusInv
   http://hgg.au.dk/software/aarhusinv/

IPI2WINLite
   http://geophys.geol.msu.ru/ipi2win.htm

JOINTEM
   https://wiki.oulu.fi/pages/viewpage.action?pageId=20677943

RES1D
   http://www.geotomosoft.com/downloads.php

VESStudy
   http://simfem-en.blogspot.fr/p/vesstudy.html

A.1.3. *Academic codes*

VES1DINV (Matlab program included in the article):

LEVENT EKINCI, Y. and DEMIRCI A., "A damped least-squares inversion program for the interpretation of Schlumberger sounding curves", *Journal of Applied Sciences*, vol. 8, no. 22, pp. 4070–4078, 2008, 10.3923/ jas.2008. 4070.4078

KOHLBECK, F. and MAWLOOD, D., "Computer program to calculate resistivities and layer thicknesses from Schlumberger soundings at the surface, at lake bottom and with two electrodes down in the subsurface", *Computers & Geosciences*, vol. 35, no. 8, pp. 1748–1571, 2009, https://doi. org/10. 1016/j.cageo.2009.03.001.

A.2. Modeling and interpreting electrical tomography images

A.2.1. *Commercial or semi-commercial software (sometimes available in demo or trial version, and even free or at a reduced price for students). 3D versions may supplement 2D software.*

AGI: EarthImager 2D
   https://www.agiusa.com/agi-earthimager-2d

Geogiga Technology Corp.: RTomo
   http://www.geogiga.com/en/rt.php

Geotomo Software: Res2dmod, Res2dinv
   http://www.geotomosoft.com/downloads.php

Geotomography: SensInv2d
   http://www.harbourdom.de/sensiv2d.htm

Multi-Phase Technologies, LLC: ERTLab64
   http://www.mpt3d.com/downloads.html

UBC Geophysical Inversion Facility: DCIP2D
   https://gif.eos.ubc.ca/software/dcip2d

ZOND: ZondRes2d
   http://zond-geo.ru/english/zond-software/ert-and-ves/zondres2d/

A.2.2. *Free, non-commercial software*

EIDORS
http://eidors3d.sourceforge.net/

Resistivity.net (Dr. Thomas Gunther): Bert, DC2dInvRes
   http://www.resistivity.net/

University of Lancaster (Pr. Andrew Binley): ProfileR, R2, R3t
   http://www.es.lancs.ac.uk/people/amb/Freeware/Freeware.htm

A.2.3. *Academic codes (codes that can generally be downloaded on the journal's website)*

IP4DI

KARAOULIS M., RÉVIL A., TSOURLOS P., WEKEMA D.D., and MINSLEY B.J., "IP4DI: A software for time-lapse 2D/3D DC-resistivity and induced polarization tomography", *Computers & Geosciences*, vol. 54, pp. 164–170, 2013, https://doi.org/10.1016/j. cageo.2013.01.008

FW2-5D

PIDLISECKY A. and KNIGHT R., "FW2_5D: A MATLAB 2.5-D electrical resistivity modeling code", *Computers & Geosciences*, vol. 34, no. 12, pp. 1645–1654, 2008, https://doi.org/10.1016/j.cageo.2008.04.001.

RESINVM3D
http://software.seg.org/2007/0001/

PIDLISECKY A., HABER E. and KNIGHT R., "RESINVM3D: A 3D resistivity inversion package", *Geophysics*, vol. 72, no. 2, pp. H1–H10, 2007. https://doi. org/10.1190/1.2402499

A.3. Links about low-tension electrical prospecting devices

Excerpts from the journal "Everyday Practical Electronics", http://www. epemag3.com/.

For the 1997 version:

    http://pe2bz.philpem.me.uk/Detect-Sense/-%20-%20Resistivity/Prj-101-Resitivity- Project-------/erm.html

    http://www.geotech1.com/cgi-bin/pages/common/index.pl?page=geo&file= projects/erm1/index.dat

For the 2003 version:

    http://www.geotech1.com/cgi-bin/pages/common/index.pl?page=geo&file=projects/erl/index.dat

Part 1: https://www.slideshare.net/jplateado/earth-resistivity-logger-john-becker

To download the journal:

    http://www.diagram.com.ua/english/library/everyday-practical-
    electronics-magazine/everyday-practical-electronics-
    magazine.php?row=5

For the 2003 version, the relevant editions are from April and May:
https://dfiles.eu/files/awvuq3xip

The program for this device can be found at:

    ftp://ftp.epemag.wimborne.co.uk/pub/PICS/EarthRes_2003

Bibliography

Further reading

BURGER H.R., SHEEHAN A.F., JONES C.H., *Introduction to Applied Geophysics: Exploring the Shallow Subsurface*, W.W. Norton & Company, 2006.

DENTITH M., MUDGE S.T., *Geophysics for the Mineral Exploration Geoscientist*, Cambridge University Press, 2014.

EVERETT M.E., *Near-Surface Applied Geophysics*, Cambridge University Press, 2013.

FLORSCH N., MUHLACH F., *Everyday Applied Geophysics 2*, ISTE Press, London and Elsevier, Oxford, 2018.

FLORSCH N., MUHLACH F., *Everyday Applied Geophysics 3*, ISTE Press, London and Elsevier, Oxford, 2018.

FROHLICH R.K., PARKE C.D., "The Electrical Resistivity of the Vadose Zone – Field Survey", *Ground Water*, vol. 27, pp. 524–530, 1989.

KIRSCH R., *Groundwater Geophysics*, Springer, 2008.

MILSON J., ERIKSEN A., *Field Geophysics*, 4th ed., Wiley, 2011.

MUALEM Y., FRIEDMAN S.P., "Theoretical Prediction of Electrical Conductivity in Saturated and Unsaturated Soil", *Water Resources Research*, vol. 27, pp. 2771–2777, 1991.

REYNOLDS J.M., *An Introduction to Applied and Environmental Geophysics*, Wiley-Blackwell, 1997.

RHOADES J., MANTEGHI N., SHROUSE P. *et al.*, "Soil Electrical Conductivity and Soil Salinity: New Formulations and Calibrations", *Soil Science Society of America Journal*, vol. 53, pp. 433–439, 1989.

RUBIN T., HUBBARD S., *Hydrogeophysics*, Springer, 2005.

TELFORD W.M., GELDART L.P., SHERIFF R.E., *Applied Geophysics*, 2nd ed., Cambridge University Press, 1990.

Index

Printed in the United States
By Bookmasters